P9-CCQ-138

Common Core Math Workouts Grade 7

AUTHORS: Karise Mace and Keegen Gennuso

EDITORS: Mary Dieterich and Sarah M. Anderson

PROOFREADER: Margaret Brown

COPYRIGHT © 2014 Mark Twain Media, Inc.

ISBN 978-1-62223-470-7

Printing No. CD-404221

Mark Twain Media, Inc., Publishers
Distributed by Carson-Dellosa Publishing LLC

Visit us at www.carsondellosa.com

Table of Contents
With Common Core State Standard Correlations

The corresponding Common Core State Standard for Mathematics is listed at the beginning of each exercise below.

Table of Contents
With Common Core State Standard Correlations (cont.)

Introduction to the Teacher

The time has come to raise the rigor in our children's mathematical education. The Common Core State Standards were developed to help guide educators and parents on how to do this by outlining what students are expected to learn throughout each grade level. The bar has been set high, but our students are up to the challenge.

This worktext is designed to help teachers and parents meet the challenges set forth by the Common Core State Standards. It is filled with skills practice and problem-solving practice exercises that correspond to each standard for mathematics. With a little time each day, your students will become better problem solvers and will acquire the skills they need to meet the mathematical expectations for their grade level.

Each page contains two "workouts." The first workout is a skills practice exercise, and the second is geared toward applying that skill to solve a problem. These workouts make great warm-up or assessment exercises. They can be used to set the stage and teach the content covered by the standards. They can also be used to assess what students have learned after the content has been taught.

We hope that this book will help you help your students build their Common Core Math strength and become great problem solvers!

Karise Mace and Keegen Gennuso

Name: _____ Date: _____

GEOMETRY – Scale Drawings

CCSS Math Content 7.G.A.1: Solve problems involving scale drawings of geometric figures, including computing actual lengths and areas from a scale drawing and reproducing a scale drawing at a different scale.

SHARPEN YOUR SKILLS:

A scale drawing of Shen's backyard is shown. Each centimeter in the drawing represents 8 feet. What is the actual area of Shen's backyard? Show your work.

_____

$10\frac{1}{2}$ cm

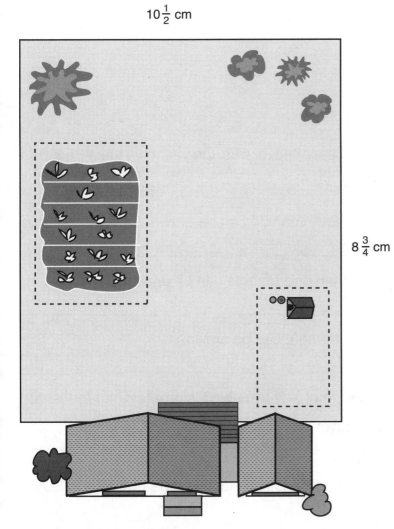

$8\frac{3}{4}$ cm

APPLY YOUR SKILLS:

On your own paper, create a new scale drawing of Shen's backyard where $\frac{1}{2}$ inch represents 7 feet. What should the length and width of the scale drawing of the backyard be? Show your work.

_____

Name: _____   Date: _____

GEOMETRY – Drawing and Constructing Geometric Shapes

CCSS Math Content 7.G.A.2: Draw (freehand, with ruler and protractor, and with technology) geometric shapes with given conditions. Focus on constructing triangles from three measures of angles or sides, noticing when the conditions determine a unique triangle, more than one triangle, or no triangle.

SHARPEN YOUR SKILLS:

Use a straightedge and protractor to draw a triangle with the given angles shown below.

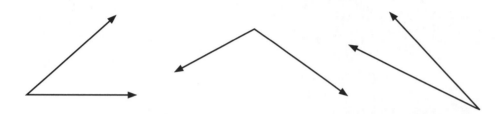

APPLY YOUR SKILLS:

Determine whether the statement is always, sometimes, or never true. On your own paper, support your answer with a sketch, and explain your reasoning.

1. Given three angles whose sum is 180°, one unique triangle can be drawn.

    _____

2. Given three angles whose sum is 180°, more than one triangle can be drawn.

    _____

3. Given three angles whose sum is not 180°, no triangle can be drawn.

    _____

4. Given three sides for which the sum of the length of any two sides is greater than the length of the third side, one unique triangle can be drawn.

    _____

5. Given three sides for which the sum of the length of any two sides is greater than the length of the third side, more than one triangle can be drawn.

    _____

6. Given three sides for which the sum of the length of two sides is less than the length of the third side, no triangle can be drawn.

    _____

Name: _____   Date: _____

GEOMETRY – Cross-Sections of Solids

SHARPEN YOUR SKILLS:

Sketch the cross-section that results from slicing the three-dimensional shape with the given plane.

Figure	Rectangular Prism	Rectangular Pyramid	Cylinder	Triangular Pyramid
Plane perpendicular to shaded base(s)				
Plane parallel to shaded base(s)				

APPLY YOUR SKILLS:

1. If a right rectangular prism is sliced by several planes that are parallel to the bases, will all of the cross-sections be congruent? Explain your reasoning.

_____

_____

_____

2. If a right rectangular pyramid is sliced by several planes that are parallel to the base, will all of the cross-sections be congruent? Explain your reasoning.

_____

_____

_____

Name: _____ Date: _____

GEOMETRY – Circumference and Area of Circles

CCSS Math Content 7.G.B.4: Know the formulas for the area and circumference of a circle and use them to solve problems; give an informal derivation of the relationship between the circumference and area of a circle.

SHARPEN YOUR SKILLS:

Calculate the circumference and area of the given circle. Show your work and leave your answers in terms of π.

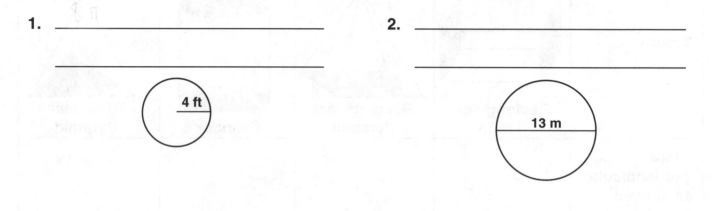

1. _____

   _____

   4 ft

2. _____

   _____

   13 m

APPLY YOUR SKILLS:

Albert is making a circular patio in his backyard with brick pavers. The patio will have a 24-foot diameter. It takes 4.5 brick pavers to cover one square foot. Approximately how many brick pavers will Albert need? Use 3.14 for π, and show your work.

_____

Name: _____   Date: _____

GEOMETRY – Angle Relationships

CCSS Math Content 7.G.B.5: Use facts about supplementary, complementary, vertical, and adjacent angles in a multi-step problem to write and solve simple equations for an unknown angle in a figure.

SHARPEN YOUR SKILLS:

1. Angle *A* and angle *B* are complementary angles. Without using a protractor, determine the measure of each angle. Show your work.

    A = _____   *B* = _____

2. Angle *C* and angle *D* are supplementary angles. Without using a protractor, determine the measure of each angle. Show your work.

    C = _____   *D* = _____

APPLY YOUR SKILLS:

In the figure below, ∠*QRS* is complementary to ∠*SRT* and supplementary to ∠*SRV*. The measure of ∠*SRV* is 136°. What is the measure of ∠*SRT*? Show how you calculated your answer.

_____

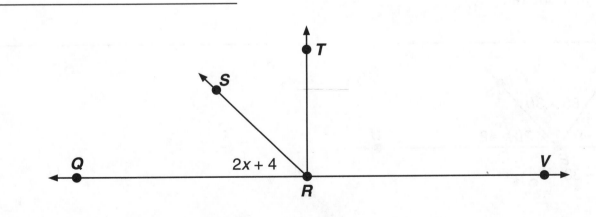

Name: _____   Date: _____

GEOMETRY – Angle Relationships

CCSS Math Content 7.G.B.5: Use facts about supplementary, complementary, vertical, and adjacent angles in a multi-step problem to write and solve simple equations for an unknown angle in a figure.

SHARPEN YOUR SKILLS:

1. In the figure below, $\angle ANG$ and $\angle LNE$ are vertical angles. Without using a protractor, determine the measures of $\angle ANG$ and $\angle LNE$. Show your work and explain how you determined your answers.

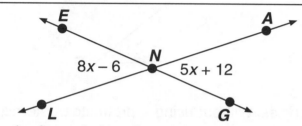

2. In the figure below, the measure of $\angle SED$ is 156° and $\angle SEI$ and $\angle IED$ are adjacent angles. Without using a protractor, determine the measures of $\angle SEI$ and $\angle IED$. Show your work and explain how you determined your answers.

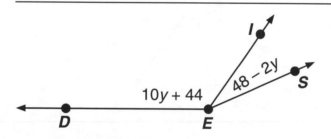

APPLY YOUR SKILLS:

Without using a protractor, determine the measure of $\angle SEU$. Show your work and explain how you determined your answer.

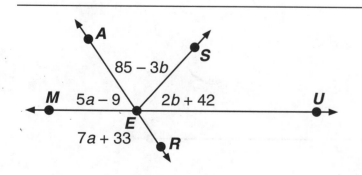

Name: _____ Date: _____

GEOMETRY – Problem Solving With Area, Volume, and Surface Area

CCSS Math Content 7.G.B.6: Solve real-world and mathematical problems involving area, volume, and surface area of two- and three-dimensional objects composed of triangles, quadrilaterals, polygons, cubes, and right prisms.

SHARPEN YOUR SKILLS:

1. Calculate the area of the isosceles triangle and parallelogram. Show your work.

    Area of Triangle = _____   Area of Parallelogram = _____

    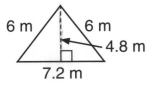

    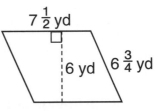

2. Calculate the surface area of the trapezoidal prism. The bases of the prism are isosceles trapezoids. Show your work.

    Surface Area of Trapezoidal Prism = _____

    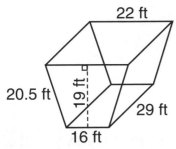

APPLY YOUR SKILLS:

Meredith is making a birthday cake. She plans to cover the top and sides of the cake with a smooth layer of fondant before decorating it. A sketch of the cake is shown at right. One batch of fondant covers 250 square inches of cake. How many batches of fondant will Meredith need to make in order to have enough to cover the cake? Will she use all of the fondant she makes? Explain how you determined your answer.

_____

_____

_____

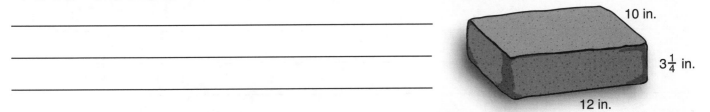

Name: _____   Date: _____

GEOMETRY – Problem Solving With Area, Volume, and Surface Area

CCSS Math Content 7.G.B.6: Solve real-world and mathematical problems involving area, volume, and surface area of two- and three-dimensional objects composed of triangles, quadrilaterals, polygons, cubes, and right prisms.

SHARPEN YOUR SKILLS:

1. Calculate the volume of the cube. Show your work.

   Volume of Cube = _____

   $8\frac{5}{8}$ ft

2. Calculate the volume of the triangular prism. Show your work.

   Volume of Triangular Prism = _____

   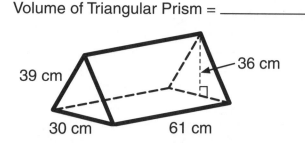
   39 cm   36 cm   30 cm   61 cm

APPLY YOUR SKILLS:

Devon built shelves in his basement to use for storage. He built the shelves so that they would easily hold plastic containers shaped like rectangular prisms that measure 16 inches by 13 inches by 1.75 feet. He currently has his off-season clothing stuffed into a box that is 1.5 feet by 1.5 feet by 14 inches. Will all of the clothing fit into a plastic storage container so that he can put it on his new shelves? Explain how you determined your answer.

_____

_____

_____

_____

_____

_____

Name: _____ Date: _____

RATIOS AND PROPORTIONAL RELATIONSHIPS –
Unit Rates

CCSS Math Content 7.RP.A.1: Compute unit rates associated with ratios of fractions, including ratios of lengths, areas, and other quantities measured in like or different terms.

SHARPEN YOUR SKILLS:

1. A recipe calls for 2 teaspoons of vanilla and $\frac{3}{4}$ cup of sugar. Compute the unit rate for vanilla to sugar. Show your work.

   _____

2. Caecilia can run $\frac{3}{20}$ of a marathon in $\frac{2}{3}$ of an hour. Compute the unit rate for marathon distance to hours. Show your work.

   _____

APPLY YOUR SKILLS:

Eric is making pizza dough. The recipe he is using will make 2 crusts and calls for $\frac{1}{4}$ cup of olive oil and $\frac{3}{8}$ ounce of yeast. Eric gets distracted while making the dough, and accidentally puts 1 ounce of yeast into the bowl with the olive oil. How much more olive oil does Eric need to add so that the dough maintains the correct ratio of olive oil and yeast? Show your work.

_____

Name: _____ Date: _____

RATIOS AND PROPORTIONAL RELATIONSHIPS –
Proportional Relationships

CCSS Math Content 7.RP.A.2a: Decide whether two quantities are in a proportional relationship, e.g., by testing for equivalent ratios in a table or graphing on a coordinate plane and observing whether the graph is a straight line through the origin.

SHARPEN YOUR SKILLS:

Determine whether or not the relationship between *x* and *y* is proportional. Explain how you determined your answer.

1.

x	y
1	7
2	14
3	21
4	28

2.

x	y
1	1
2	4
3	9
4	16

3.

x	y
−2	−1
2	1
4	2
6	3

APPLY YOUR SKILLS:

1. Would you expect the relationship between the number of 50-meter laps Trevor swims and the time it takes him to swim those laps to be proportional? Explain your reasoning.

_____

_____

2. Trevor records the time it takes him to swim 50-meter laps in the table. Are these two quantities proportional? Explain your reasoning.

Number of Laps	Time (in minutes)
1	1
2	1.9
5	6.5
10	12.2

Name: _____   Date: _____

RATIOS AND PROPORTIONAL RELATIONSHIPS –
Proportional Relationships

CCSS Math Content 7.RP.A.2a: Decide whether two quantities are in a proportional relationship, e.g., by testing for equivalent ratios in a table or graphing on a coordinate plane and observing whether the graph is a straight line through the origin.

SHARPEN YOUR SKILLS:

Determine whether or not the graph shows relationships that are proportional. Explain how you determined your answer.

1.   **2.**   **3.**

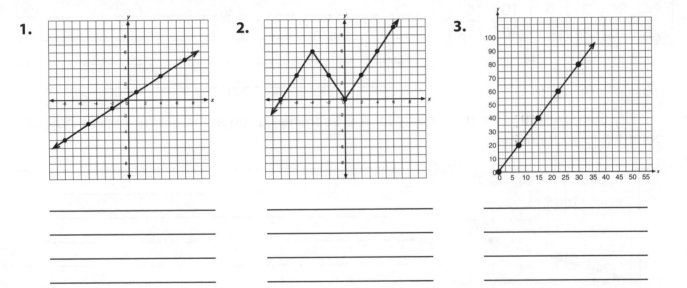

_____   _____   _____

_____   _____   _____

_____   _____   _____

_____   _____   _____

APPLY YOUR SKILLS:

The graph shows the growth of a stock over time. Use the graph to complete the following exercises.

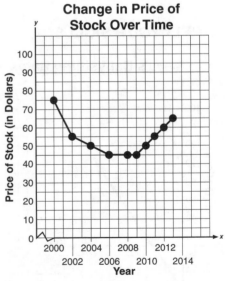

Change in Price of Stock Over Time

1. Is the relationship between the price of the stock and time proportional? Explain your reasoning.

   _____

   _____

   _____

2. What would the price of the stock need to be in 2014 so that the relationship between stock price and time is proportional between the years 2009 and 2014? Explain how you determined your answer.

_____

_____

_____

Name: _____ Date: _____

RATIOS AND PROPORTIONAL RELATIONSHIPS –
Constant of Proportionality

CCSS Math Content 7.RP.A.2b: Identify the constant of proportionality (unit rate) in tables, graphs, equations, diagrams, and verbal descriptions of proportional relationships.

SHARPEN YOUR SKILLS:

1. Identify the constant of proportionality for the quantities given in the table. Explain how you determined your answer.

x	2	3	6	10	14
y	16	24	48	80	112

_____

_____

_____

_____

2. Identify the constant of proportionality for the quantities given in the graph. Explain how you determined your answer.

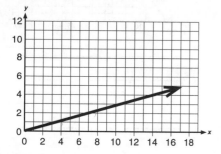

_____

_____

_____

_____

_____

APPLY YOUR SKILLS:

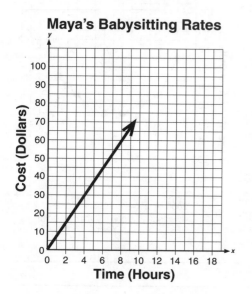

Maya's Babysitting Rates

Use the graph to determine the amount Maya charges to babysit for one hour. Explain how you determined your answer.

_____

_____

_____

_____

Name: _____ Date: _____

RATIOS AND PROPORTIONAL RELATIONSHIPS –
Constant of Proportionality

CCSS Math Content 7.RP.A.2b: Identify the constant of proportionality (unit rate) in tables, graphs, equations, diagrams, and verbal descriptions of proportional relationships.

SHARPEN YOUR SKILLS:

Identify the constant of proportionality in the given equation.

1. $y = 2x$ _____

2. $y = \frac{5}{3}x$ _____

3. A 5-pound bag of apples costs $4.80. What is the price for one pound of apples? Show your work.

    _____

4. A crayon factory can make about 35 million crayons in 7 days. How many crayons can the factory make in one day? Show your work.

    _____

APPLY YOUR SKILLS:

A $2\frac{1}{2}$-pound bag of white grapes costs $4.70. A $3\frac{1}{4}$-pound bag of red grapes costs $5.98. Which bag of grapes is a better deal? Explain how you determined your answer.

_____

_____

_____

_____

Name: _____ Date: _____

RATIOS AND PROPORTIONAL RELATIONSHIPS –
Proportional Relationships

CCSS Math Content 7.RP.A.2c: Represent proportional relationships by equations.

SHARPEN YOUR SKILLS:

Jody is observing fireflies. He catches one and places it in a jar. Then he counts the number of times it flashes in one minute and two minutes. Once he has recorded his observations, Jody releases the firefly, catches another, and counts the number of times it flashes in one and two minutes. He continues this process until he has observed 6 different fireflies. The tables show the relationship between time and the number of flashes for each of Jody's fireflies.

Table 1

Firefly	A	B	C	D	E	F
Time (minutes)	1	1	1	1	1	1
Number of flashes	15	13	18	19	12	14

Table 2

Firefly	A	B	C	D	E	F
Time (minutes)	2	2	2	2	2	2
Number of flashes	28	26	32	30	25	29

1. Write an equation that represents the relationship between time, t, and the number of flashes, f, for each of the fireflies. Use the data from Table 1.

   _____

   _____

2. Write an equation that represents the relationship between time, t, and the number of flashes, f, for each of the fireflies. Use the data from Table 2.

   _____

   _____

APPLY YOUR SKILLS:

A recipe for granola calls for 2 cups of raisins and 3 cups of sunflower seeds. Mr. Stahler asks his students to write an equation that represents the relationship between raisins and sunflower seeds. Two students' equations are shown below. Do these equations represent the same relationship? Explain how you determined your answer.

   Student 1: $s = \frac{3}{2}r$ **Student 2:** $r = \frac{2}{3}s$

_____

_____

_____

_____

Name: _____ Date: _____

RATIOS AND PROPORTIONAL RELATIONSHIPS –
Proportional Relationships and Graphs

CCSS Math Content 7.RP.A.2d: Explain what a point (*x*, *y*) on the graph of a proportional relationship means in terms of the situation, with special attention to the points (0, 0) and (1, *r*) where *r* is the unit rate.

SHARPEN YOUR SKILLS:

The graph shows the relationship between time and the distance Darrell has ridden on his bicycle. Use the graph to complete the exercises.

1. Explain what point *A* represents on the graph.

_____

_____

_____

_____

2. Explain what the point (0, 0) represents on the graph.

_____

_____

_____

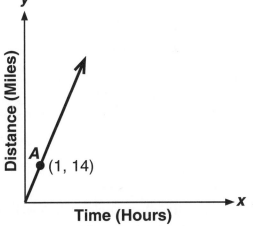

APPLY YOUR SKILLS:

Luther sold corn at the local farmer's market at a rate of $5.28 for 3 dozen ears of corn. Which of the points on the graph below corresponds to the rate at which Luther sold the corn? Explain how you determined your answer.

_____

_____

_____

_____

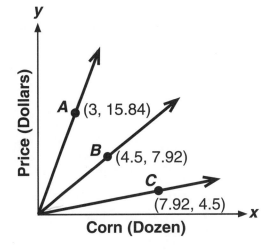

Name: _____ Date: _____

RATIOS AND PROPORTIONAL RELATIONSHIPS –
Problem Solving With Proportional Relationships

CCSS Math Content 7.RP.A.3: Use proportional relationships to solve multi-step ratio and percent problems.

SHARPEN YOUR SKILLS:

Margaret is selling her house. Her original asking price was $372,000. She recently dropped her asking price to $327,360. Calculate the percent decrease in Margaret's asking price for her house. Show your work.

Percentage decrease = _____

APPLY YOUR SKILLS:

River collects baseball cards and hats. He has been saving his money so that he can add to his collection this summer. A pack of baseball cards costs $2.75, and a baseball hat costs $29.95. He really wants to expand his baseball card collection, so he decides to buy 6 times as many packs of cards as hats. The sales tax in River's state is 7.5%. River has $200 to spend. Determine how many packs of baseball cards and baseball hats River can buy. Show your work.

Name: _____   Date: _____

THE NUMBER SYSTEM – Opposite Quantities With Zero Sum

CCSS Math Content 7.NS.A.1a: Describe situations in which opposite quantities combine to make 0.

SHARPEN YOUR SKILLS:

Determine whether or not the situation is one in which opposite quantities combine to make 0. Explain how you determine your answer.

1. Parker's account balance was –$12.84. So he made a deposit of $20.00.

   _____

   _____

   _____

2. The football team lost 10 yards on their first down. Then they gained 10 yards on their second down. _____

   _____

   _____

3. Kristen started her hike at an elevation of 345 feet above sea level. Then she hiked down into a canyon and stopped when she was 500 feet below sea level.

   _____

   _____

   _____

4. Aiden had an account balance of $200. Then he withdrew $200 to buy a new bike.

   _____

   _____

   _____

APPLY YOUR SKILLS:

Describe two different situations in which opposite quantities combine to make 0.

_____

_____

_____

_____

_____

_____

Name: _____ Date: _____

THE NUMBER SYSTEM –
Understanding Addition of Rational Numbers

CCSS Math Content 7.NS.A.1b: Understand $p + q$ as the number located a distance $|q|$ from p, in the positive or negative direction depending on whether q is positive or negative. Show that a number and its opposite have a sum of 0 (additive inverses). Interpret sums of rational numbers by describing real-world contexts.

SHARPEN YOUR SKILLS:

Determine the point that represents the sum given in the exercise. Explain how you determined your answer.

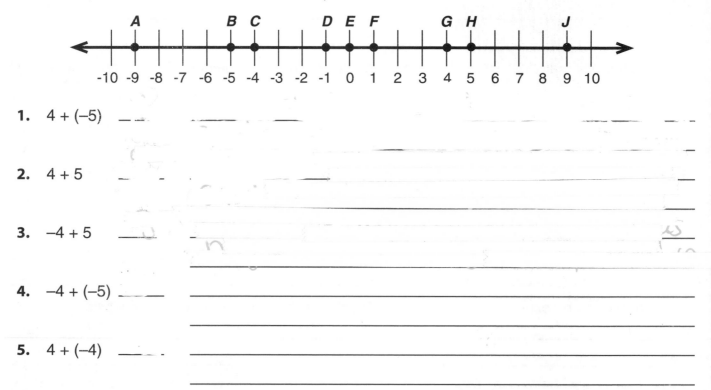

1. 4 + (–5) ___ _ _____ ___ ___

_____

2. 4 + 5 __ _ _ _____ __

_____

3. –4 + 5 ___ _ _

_____

4. –4 + (–5) _ _ _ _____

_____

5. 4 + (–4) ___ _ _ _____

_____

APPLY YOUR SKILLS:

The temperature at 7:00 A.M. was –8°F. By noon, the temperature had risen 6 degrees.

1.   Write an equation that represents this situation. _____

2.   Determine the temperature at noon. Explain how you determined your answer. _____

_____

_____

_____

Name: _____   Date: _____

THE NUMBER SYSTEM –
Understanding Subtraction of Rational Numbers

CCSS Math Content 7.NS.A.1c: Understand subtraction of rational numbers as adding the additive inverse, $p - q = p + (-q)$. Show that the distance between two rational numbers on the number line is the absolute value of their difference, and apply this principle in real-world contexts.

SHARPEN YOUR SKILLS:

Rewrite the subtraction expression as an addition expression.

1. $8 - 3$

___

4. $-2 - 7$

___

2. $9.2 - (-4.9)$

___

5. $4 - 12.3$

___

3. $-\frac{3}{5} - \frac{2}{5}$

___

6. $\frac{2}{9} - \frac{5}{6}$

APPLY YOUR SKILLS:

The thermometer registered a temperature of –2°C at 8 P.M. By midnight, the temperature had dropped to –9°C. How much did the temperature drop between 8 P.M. and midnight? Explain how you determined your answer. Support your explanation with a number line diagram.

_____

_____

_____

_____

_____

Name: _____ Date: _____

THE NUMBER SYSTEM –
Adding and Subtracting Rational Numbers

CCSS Math Content 7.NS.A.1d: Apply properties of operations as strategies to add and subtract rational numbers.

SHARPEN YOUR SKILLS:

Calculate the sum.

1. $15 + (-47)$

_____ __

3. $-28.73 + (-42.21)$

_____

2. $-\frac{4}{7} + \frac{6}{7}$

_____ __

4. $\frac{4}{15} + \frac{7}{20}$

_____

APPLY YOUR SKILLS:

1. Danielle has worked 5.3 hours so far this week. For the remainder of the week, she is scheduled to work $22\frac{1}{2}$ more hours. How many total hours will Danielle work this week? Show your work.

_____

_____

_____

2. Dave dug a hole for a tree he was planting that was 2.5 feet deep. The tree was 7.4 feet tall from roots to top. What was the height of the tree above ground after it was planted? Show your work.

_____

_____

_____

Name: _____   Date: _____

THE NUMBER SYSTEM –
Adding and Subtracting Rational Numbers

CCSS Math Content 7.NS.A.1d: Apply properties of operations as strategies to add and subtract rational numbers.

SHARPEN YOUR SKILLS:

Calculate the difference.

1. $13.7 - 86.9$

_____

3. $58 - 17\frac{3}{8}$

_____

2. $-\frac{5}{9} - \left(-\frac{8}{9}\right)$

_____

4. $-161 - 122$

_____

APPLY YOUR SKILLS:

1. Jacqueline is $1\frac{7}{10}$ meters tall and Roderick is 1.88 meters tall. How much taller is Roderick? Show your work.

_____

_____

_____

2. Sarah got on the elevator outside her office on the 17th floor. She went down to the records department, which is located 9 floors below ground level. How many floors down did she go in total? Show your work.

_____

_____

_____

Name: _____   Date: _____

THE NUMBER SYSTEM –
Understanding Multiplication of Rational Numbers

CCSS Math Content 7.NS.A.2a: Understand that multiplication is extended from fractions to rational numbers by requiring that operations continue to satisfy the properties of operations, particularly the distributive property, leading to products such as $(-1)(-1) = 1$ and the rules for multiplying signed numbers. Interpret products of rational numbers by describing real-world contexts.

SHARPEN YOUR SKILLS:

Use the properties of multiplication to rewrite the expression.

1. $\left(\frac{2}{5} \times \left(-\frac{5}{8}\right)\right) \times \left(-\frac{4}{7}\right)$ _____

_____

2. $-8((-1.5) + 4.25)$ _____

_____

APPLY YOUR SKILLS:

Christopher's elevation decreased by $\frac{4}{5}$ mile every hour of his 3-hour hike. Write an expression that could be used to determine the total change in elevation during his hike. Explain how you determined your answer.

_____

_____

_____

_____

_____

_____

Name: _____ Date: _____

THE NUMBER SYSTEM –
Understanding Division of Rational Numbers

CCSS Math Content 7.NS.A.2b: Understand that integers can be divided, provided that the divisor is not zero, and every quotient of integers (with non-zero divisor) is a rational number. If p and q are integers, then $-(p/q) = (-p)/q = p/(-q)$. Interpret quotients of rational numbers by describing real-world contexts.

SHARPEN YOUR SKILLS:

Write two other expressions that are equivalent to the given expression.

1. $-\left(\dfrac{9}{7}\right)$ _____   _____

2. $\dfrac{(-5)}{8}$ _____   _____

3. $\dfrac{12}{(-5)}$ _____   _____

APPLY YOUR SKILLS:

Tim's company lost $32,857 over the course of 6 months. Write an expression that could be used to estimate how much money Tim's company lost each month. Explain how you determined your answer.

_____

_____

_____

_____

_____

Name: _____   Date: _____

THE NUMBER SYSTEM –
Multiplying and Dividing Rational Numbers

CCSS Math Content 7.NS.A.2c: Apply properties of operations as strategies to multiply and divide rational numbers.

SHARPEN YOUR SKILLS:

Calculate the product. Give your answer in simplest form.

1. $-18 \times (-4)$

_____

3. $2.45 \times (-12.3)$

_____

2. $-\dfrac{4}{9} \times \dfrac{27}{32}$

_____

4. $\dfrac{4}{15} \times \dfrac{7}{15}$

_____

APPLY YOUR SKILLS:

1. A submarine is descending at a rate of 182.25 feet per minute. How far below sea level will the submarine be after $6\frac{1}{2}$ minutes? Show your work.

_____

_____

_____

2. Katie's Pie Place sells an average of 6 pieces of pie per hour. Last week Katie's shop was open 37.5 hours. How many pieces of pie did she sell last week? Show your work.

_____

_____

_____

Name: _____   Date: _____

THE NUMBER SYSTEM –
Multiplying and Dividing Rational Numbers

CCSS Math Content 7.NS.A.2c: Apply properties of operations as strategies to multiply and divide rational numbers.

SHARPEN YOUR SKILLS:

Calculate the quotient. Give your answer in simplest form.

1. $\frac{4}{5} \div \left(-\frac{8}{9}\right)$

    _____

3. $-\frac{18}{35} \div \frac{3}{7}$

    _____

2. $-475 \div (-19)$

    _____

4. $314.55 \div (-13.98)$

    _____

APPLY YOUR SKILLS:

1. A coal mining company is installing a new elevator down into the mine. The elevator will go –3,280.6 feet into the ground. It will make stops every 500 feet. How many stops will the elevator make before it reaches its greatest depth? Show your work.

    _____

    _____

    _____

2. Last week, the mining company brought up 94,820 tons of coal. If the company was running for six days, how many tons of coal were brought up each day? Round your answer to the nearest hundredth. Show your work.

    _____

    _____

    _____

Name: _____ Date: _____

THE NUMBER SYSTEM –
Writing Rational Numbers as Decimals

CCSS Math Content 7.NS.A.2d: Convert a rational number to a decimal using long division; know that the decimal form of a rational number terminates in 0s or eventually repeats.

SHARPEN YOUR SKILLS:

Convert the rational number to a decimal using long division.

1. $\frac{15}{22}$ **2.** $\frac{5}{8}$ **3.** $\frac{643}{500}$

APPLY YOUR SKILLS:

The number π (pi) is not a rational number. The fraction $\frac{22}{7}$ is sometimes used to approximate π. Eva argues that if π is not rational, then $\frac{22}{7}$ must be not be rational either. Is she correct? Explain how you determined your answer and support your explanation with mathematics.

_____

_____

_____

_____

_____

_____

Name: _____   Date: _____

THE NUMBER SYSTEM –
Problem Solving With Rational Numbers

CCSS Math Content 7.NS.A.3: Solve real-world and mathematical problems involving the four operations with rational numbers.

SHARPEN YOUR SKILLS:

1. What number is $\frac{5}{8}$ more than $\frac{3}{16}$? Show how you calculated your answer.

   _____

   _____

   _____

2. The public library is $\frac{9}{10}$ of a mile from the school. The community center is $\frac{1}{4}$ of a mile from the school. How much farther from the school is the library than the community center? Show how you calculated your answer.

   _____

   _____

   _____

APPLY YOUR SKILLS:

Jan has a curtain that is $1\frac{3}{4}$ yards long. She wants to increase the length of the curtain by $\frac{1}{3}$ of a yard. Jan has $\frac{7}{8}$ of a yard of lace. She plans to cut a piece from the lace and sew it onto the bottom of the curtain.

1. How long will the curtain be after Jan adds the lace? Show your work.

   _____

   _____

2. How much lace will be left after Jan cuts the piece off to add to the curtain? Show your work.

   _____

   _____

   _____

Name: _____ Date: _____

THE NUMBER SYSTEM –
Problem Solving With Rational Numbers

CCSS Math Content 7.NS.A.3: Solve real-world and mathematical problems involving the four operations with rational numbers.

SHARPEN YOUR SKILLS:

1. What number is $\frac{2}{3}$ of $\frac{9}{14}$? Show how you calculated your answer.

   _____

   _____

   _____

2. Steven has $\frac{3}{4}$ of a ton of topsoil in his dump truck. He needs to divide it into $\frac{1}{8}$-ton piles. How many piles will Steven have? Show how you calculated your answer.

   _____

   _____

   _____

APPLY YOUR SKILLS:

DeWayne has a garden that is $18\frac{1}{2}$ feet by $12\frac{1}{4}$ feet.

1. How many square feet is DeWayne's garden? Show how you calculated your answer.

   _____

   _____

   _____

2. If DeWayne divides his garden into rectangles that are $3\frac{5}{8}$ feet by $1\frac{2}{3}$ feet, how many rectangles will he have? Show how you calculated your answer.

   _____

   _____

   _____

   _____

Name: _____ Date: _____

EXPRESSIONS AND EQUATIONS –
Manipulating Linear Expressions

CCSS Math Content 7.EE.A.1: Apply properties of operations as strategies to add, subtract, factor, and expand linear expressions with rational coefficients.

SHARPEN YOUR SKILLS:

Add or subtract as indicated.

1. $\left(\frac{2}{7}x + 8\right) + \left(\frac{3}{7}x - 4\right)$

2. $\left(\frac{11}{15}a - 21\right) - \left(\frac{2}{5}a + 16\right)$

APPLY YOUR SKILLS:

Mr. Osbourne wrote the following expression on the board and asked his students to simplify it.

$$\left(\frac{7}{12}y + 4\right) - \left(\frac{5}{6}y - 11\right) + \left(\frac{3}{4}y - 9\right)$$

The work of one student is shown below. Check the student's work to determine whether or not the expression is simplified correctly. If there is a mistake, identify and correct the mistake.

$$\left(\frac{7}{12}y + 4\right) - \left(\frac{5}{6}y - 11\right) + \left(\frac{3}{4}y - 9\right) = \frac{7}{12}y + 4 - \frac{10}{12}y - 11 + \frac{9}{12}y - 9$$

$$= \frac{7}{12}y - \frac{10}{12}y + \frac{9}{12}y + 4 - 11 - 9$$

$$= \frac{6}{12}y - 16$$

$$= \frac{1}{2}y - 16$$

Name: _____ Date: _____

EXPRESSIONS AND EQUATIONS –
Manipulating Linear Expressions

CCSS Math Content 7.EE.A.1: Apply properties of operations as strategies to add, subtract, factor, and expand linear expressions with rational coefficients.

SHARPEN YOUR SKILLS:

1. Factor the expression $\frac{5}{24}x + \frac{10}{36}$ completely. _____

2. Expand the expression $\frac{4}{9}\left(\frac{2}{5}b - \frac{3}{8}\right)$. _____

3. Factor the expression $\frac{6}{45} - \frac{8}{63}y$ completely. _____

4. Expand the expression $-\frac{2}{3}\left(\frac{8}{9} - \frac{2}{5}a\right)$. _____

APPLY YOUR SKILLS:

Is there more than one way to factor the expression $\frac{24}{32}a + \frac{48}{56}$? Support your answer with mathematics.

_____

_____

_____

_____

_____

Name: _____   Date: _____

EXPRESSIONS AND EQUATIONS –
Rewriting Linear Expressions

CCSS Math Content 7.EE.A.2: Understand that rewriting an expression in different forms in a problem context can shed light on the problem and how the quantities in it are related.

SHARPEN YOUR SKILLS:

Are the following expressions equivalent? If so, explain how you determined your answer. If not, explain how they differ.

Expression #1: $\frac{2}{5}(4r) - \frac{3}{10}r$

Expression #2: $13r$

_____

_____

_____

_____

APPLY YOUR SKILLS:

Two different clients are bidding on the same house. Client #2 bids 8% more than client #1. The expression $x + 0.08x$ can be used to determine the amount that client #2 bid. Write an expression that is equivalent to the given expression, and explain how it might be more helpful in determining the amount that client #2 bid.

_____

_____

_____

_____

_____

Name: _____ Date: _____

EXPRESSIONS AND EQUATIONS –
Problem Solving With Rational Numbers

CCSS Math Content 7.EE.B.3: Solve multi-step real-life and mathematical problems posed with positive and negative rational numbers in any form (whole numbers, fractions, and decimals), using tools strategically. Apply properties of operations to calculate with numbers in any form; convert between forms as appropriate; and assess the reasonableness of answers using mental computation and estimation strategies.

SHARPEN YOUR SKILLS:

Mrs. Teurk wants to purchase paper to cover her bulletin board. The paper comes on a roll that is 54 inches wide and costs $0.25 for each linear foot. Mrs. Teurk's bulletin board is $7\frac{3}{4}$ feet wide and 4.5 feet high. The office supply store also charges a $1.23 cutting fee. Simplify the following expression to see how much it will cost to purchase enough paper to cover Mrs. Teurk's bulletin board.

$$\text{Cost} = 0.25\left(7\tfrac{3}{4}\right) + 1.23$$

_____

_____

APPLY YOUR SKILLS:

A mother blue whale weans her calf when it is 8 months old. At this age, the calf weighs about $23\frac{3}{8}$ tons. On the average, blue whale calves gain about 200 pounds per day. If the calf grows at the given rate, how much will it weigh when it is 1.5 years old? Show your work and give your answer in tons.

_____

_____

_____

_____

Name: _____   Date: _____

EXPRESSIONS AND EQUATIONS –
Problem Solving With Equations

CCSS Math Content 7.EE.B.4a: Solve word problems leading to equations of the form $px + q = r$ and $p(x + q) = r$, where p, q, and r are specific rational numbers. Solve equations of these forms fluently. Compare an algebraic solution to an arithmetic solution, identifying the sequence of operations used in each approach.

SHARPEN YOUR SKILLS:

Solve the equation.

1. $\frac{1}{5}x + 7 = 16$

3. $-\frac{5}{9}a - \frac{2}{3} = \frac{5}{6}$

2. $2.4y - 9.5 = 4.66$

4. $3.25 + 1.58b = 12.3982$

APPLY YOUR SKILLS:

Lois is making applesauce. Each batch she makes yields $3\frac{1}{4}$ quarts of apple-sauce. Lois already has 1.5 quarts. She would like to have a total of $14\frac{1}{2}$ quarts before she's finished for the day. How many batches of applesauce does Lois need to make? Show your work.

_____

_____

_____

_____

_____

Name: _____ Date: _____

EXPRESSIONS AND EQUATIONS –
Problem Solving With Equations

CCSS Math Content 7.EE.B.4a: Solve word problems leading to equations of the form $px + q = r$ and $p(x + q) = r$, where p, q, and r are specific rational numbers. Solve equations of these forms fluently. Compare an algebraic solution to an arithmetic solution, identifying the sequence of operations used in each approach.

SHARPEN YOUR SKILLS:

Solve the equation. Show your work.

1. $\frac{2}{3}(x + 7) = 5$ **2.** $1.9(x - 2.7) = 12.16$

APPLY YOUR SKILLS:

Mr. Broderick presents the following problem to his students.

Danica is $1\frac{1}{2}$ years older than Evan. Jeremy is $36\frac{7}{8}$, which is 2.5 times as old as Danica. How old is Evan?

Student #1 decides to solve the problem algebraically. Student #2 decides to solve the problem arithmetically. Part of each of their solutions is shown below. Complete their solutions and then write a sentence or two comparing their solutions.

Student #1: Let e represent Evan's age. $2.5\left(e + 1\frac{1}{2}\right) = 36\frac{7}{8}$

Student #2

Evan's Age	Danica's Age	Jeremy's Age	Is Jeremy's Age $36\frac{7}{8}$?
10	11.5	28.75	No

_____

_____

_____

Name: _____ Date: _____

EXPRESSIONS AND EQUATIONS –
Problem Solving With Inequalities

CCSS Math Content 7.EE.B.4b: Solve word problems leading to inequalities of the form $px + q > r$ and $px + q < r$, where p, q, and r are specific rational numbers. Graph the solution set of the inequality and interpret it in the context of the problem.

SHARPEN YOUR SKILLS:

Solve the inequality and graph its solution set on the number line.

1. $\frac{8}{15}x + \frac{4}{5} > \frac{29}{30}$

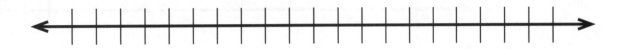

2. $5.85r - 2.93 > 25.033$

APPLY YOUR SKILLS:

Hannah has $482.75 in her savings account. She would like her account balance to be more than $1,250.00 before she buys a cell phone. Hannah plans to save $85.25 a month. How many months will it be until she can buy a cell phone? Write and solve an inequality that represents this situation. Then graph the solution set on the number line and explain what it means in the context of the situation.

_____

_____

_____

_____

_____

Name: _____ Date: _____

EXPRESSIONS AND EQUATIONS –
Problem Solving With Inequalities

CCSS Math Content 7.EE.B.4b: Solve word problems leading to inequalities of the form $px + q > r$ and $px + q < r$, where p, q, and r are specific rational numbers. Graph the solution set of the inequality and interpret it in the context of the problem.

SHARPEN YOUR SKILLS:

Solve the inequality and graph its solution set on the number line.

1. $15.46r + 28.9 < 240.702$

2. $\frac{7}{9}g + \frac{2}{3} < \frac{14}{15}$

APPLY YOUR SKILLS:

Erin is tracking the change in the tide on the beach. The tide is currently going out and has already receded $1\frac{3}{5}$ feet. It appears to be receding at a rate of approximately $\frac{19}{25}$ foot each hour. Based on her research, Erin anticipates that it will recede a total of $9\frac{1}{5}$ feet before it starts coming back in.

She needs to leave the beach for a while, but wants to return in time to see the tide at its lowest. How many hours can Erin spend away from the beach without missing seeing the tide at its lowest? Write and solve an inequality that represents this situation. Then graph the solution set on the number line and explain what it means in the context of the situation.

_____

_____

_____

_____

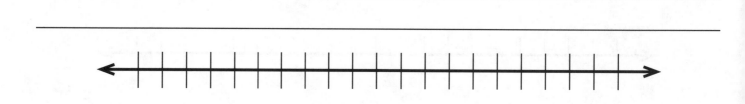

Name: _____   Date: _____

STATISTICS AND PROBABILITY –
Using Statistics to Understand Populations

CCSS Math Content 7.SP.A.1: Understand that statistics can be used to gain information about a population by examining a sample of the population; generalizations about a population from a sample are valid only if the sample is representative of that population. Understand that random sampling tends to produce representative samples and support valid inferences.

SHARPEN YOUR SKILLS:

1. A marketing person for a company that manufactures roller coasters reports that 92% of Americans like to ride roller coasters. She based this statistic on the data she collected from a survey that was given to people who visited American amusement parks over the course of the summer. Is this statistic valid? Explain your reasoning.

_____

_____

_____

2. The school newspaper reports that 47% of seventh graders at the school have at least two siblings. This statistic is based on the data a seventh grader collected from his 3rd period class. Is this statistic valid? Explain your reasoning.

_____

_____

_____

APPLY YOUR SKILLS:

Explain how the people in exercises 1 and 2 above could have collected the data so that it would be representative of the population they indicated in their report.

_____

_____

_____

_____

Name: _____   Date: _____

STATISTICS AND PROBABILITY –
Using Data to Draw Inferences

CCSS Math Content 7.SP.A.2: Use data from a random sample to draw inferences about a population with an unknown characteristic of interest. Generate multiple samples (or simulated samples) of the same size to gauge the variation in estimates or predictions.

SHARPEN YOUR SKILLS:

Jake stocked his pond with bluegill and largemouth bass. He knows that he put 120 fish in the pond, but he cannot remember how many of each type. He uses a net to take several samples of the fish. After he takes each sample, he returns it to the pond before collecting the next sample. His results are given in the table below.

Sample	1	2	3	4	5
Number of Bluegill	5	6	7	4	2
Number of Largemouth Bass	3	3	3	2	1

Based on the data Jake collected, how many of each type of fish would you predict are in the pond? Explain how you determined your answers.

_____

_____

_____

_____

APPLY YOUR SKILLS:

Darlene works in quality control at a blue-jean factory. Every hour, she randomly selects 8 pairs of blue jeans and inspects them for flaws. If more than 15% of the jeans she selects are flawed on average over the course of the day, then Darlene must inspect the machines and makes adjustments so that they will no longer produce those flaws. The results of her inspections on two days are shown below. On which day did Darlene have to make adjustments to the machines? Explain how you determined your answer.

Day 1

Sample	1	2	3	4	5	6	7	8
Number of flawed pairs of jeans	1	3	0	2	1	1	0	1

Day 2

Sample	1	2	3	4	5	6	7	8
Number of flawed pairs of jeans	3	3	2	0	0	0	4	0

_____

_____

_____

_____

Name: _____ Date: _____

STATISTICS AND PROBABILITY –
Comparing Data Sets With Similar Variabilities

CCSS Math Content 7.SP.B.3: Informally assess the degree of visual overlap of two numerical data distributions with similar variabilities, measuring the difference between the centers by expressing it as a multiple of a measure of variability.

SHARPEN YOUR SKILLS:

The double dot plot shows the heights of the members of the women's and men's basketball teams.

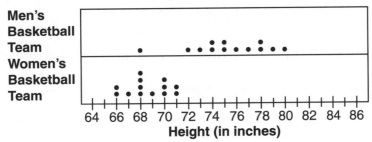

1. Which team appears to have the greater average height? Explain how you determined your answer.

_____

_____

_____

2. Which team appears to have greater variability in the heights? Explain how you determined your answer.

_____

_____

_____

APPLY YOUR SKILLS:

The data in the tables below represents the grades on the first and second quizzes in Ms. Cole's math class. On your own paper, make a double dot plot of the data and then write a few sentences comparing the grades on the quizzes.

Quiz 1

76	75	76	52	88	91	92	95	69	72
88	89	63	90	74	88	88	88	75	95

Quiz 2

71	74	80	77	77	64	68	65	65	77
67	75	77	76	76	77	74	72	70	70

Name: _____ Date: _____

STATISTICS AND PROBABILITY – Using Statistics to Infer

CCSS Math Content 7.SP.B.4: Use measures of center and measures of variability for numerical data from random samples to draw informal comparative inferences about two populations.

SHARPEN YOUR SKILLS:

The data in the table below is a random sample of the hourly pay rate of 10 working people in two different neighborhoods.

Neighborhood #1	$7.25	$11.50	$7.80	$8.10	$8.45	$7.95	$7.50	$8.75	$9.02	$8.30
Neighborhood #2	$12.00	$7.80	$11.95	$10.45	$10.08	$9.65	$10.50	$10.15	$9.75	$12.25

1. Calculate the mean of each data set. Round your answers to the nearest tenth.

    _____

    _____

2. Use your answer from exercise 1 to calculate the mean absolute deviation of each data set. Round your answers to the nearest tenth.

    _____

    _____

3. Calculate the median of each data set. _____

    _____

4. Calculate the interquartile range of each data set. _____

    _____

APPLY YOUR SKILLS:

Write a short paragraph comparing the populations of the two different neighborhoods above. Use the statistics you calculated in the exercises above.

_____

_____

_____

_____

_____

_____

_____

Name: _____   Date: _____

STATISTICS AND PROBABILITY – Understanding Probability

CCSS Math Content 7.SP.C.5: Understand that the probability of a chance event is a number between 0 and 1 that expresses the likelihood of the event occurring. Larger numbers indicate greater likelihood. A probability near 0 indicates an unlikely event, a probability near 1/2 indicates an event that is neither unlikely nor likely, and a probability near 1 indicates a likely event.

SHARPEN YOUR SKILLS:

$$0.1 \qquad \frac{1}{3} \qquad \frac{10}{11} \qquad 0.05 \quad 0.99 \quad 0.45 \quad 1\frac{2}{5} \quad \frac{7}{15} \quad 3.5 \quad 0.53$$

1. Which of the numbers above could represent the probability of an unlikely event? Explain how you determined your answer.

   _____

   _____

2. Which of the numbers above could represent the probability of an event that is neither un-likely nor likely? Explain how you determined your answer.

   _____

   _____

3. Which of the numbers above could represent the probability of a likely event? Explain how you determined your answer.

   _____

   _____

4. Which of the numbers above could not be probabilities? Explain how you determined your answer.

   _____

   _____

APPLY YOUR SKILLS:

Give 3 examples of probabilities for the type of event. Do not use the numbers from the exercises above.

1. Likely event _____

   _____

2. Unlikely event _____

   _____

3. Neither likely nor unlikely _____

   _____

Name: _____   Date: _____

STATISTICS AND PROBABILITY – Using Experimental Probability to Understand Theoretical Probability

CCSS Math Content 7.SP.C.6: Approximate the probability of a chance event by collecting data on the chance process that produces it and observing its long-run relative frequency, and predict the approximate relative frequency given the probability.

SHARPEN YOUR SKILLS:

A six-sided number cube is rolled.

1. Predict the probability of rolling a 4. Explain how you determined your answer.

   _____
   _____

2. Roll a six-sided number cube 18 times and record your results for each roll.

   _____
   _____

3. How many times did you roll a 4 in exercise 2?

   _____
   _____

4. Based on your answer to exercise 3, would you change your answer to exercise 1? Explain your reasoning.

   _____
   _____

APPLY YOUR SKILLS:

1. The probability of a fair coin landing on heads when it is flipped is $\frac{1}{2}$. If a coin is flipped 64 times, how many times would you expect it to land on heads? Explain how you determined your answer.

   _____

2. The probability of drawing a heart from a standard deck of playing cards is $\frac{1}{4}$. If you randomly draw from a deck of cards 100 times and replace the card each time, how many times would you expect to draw a heart? Explain how you determined your answer.

   _____

3. The probability of a certain spinner landing on yellow is $\frac{2}{3}$. If you spin this spinner 45 times, how many times would you expect the spinner to land on yellow? Explain how you determined your answer.

   _____

Name: _____ Date: _____

STATISTICS AND PROBABILITY –
Uniform Probability Models

CCSS Math Content 7.SP.C.7a: Develop a uniform probability model by assigning equal probability to all outcomes, and use the model to determine probabilities of events.

SHARPEN YOUR SKILLS:

There are 10 red marbles, 10 yellow marbles, 10 green marbles, 10 blue marbles, 10 black marbles, and 10 white marbles in a bag. Which probability model should you use to determine the probability of drawing a yellow marble from the bag – a six-sided number cube or a fair coin? Explain your reasoning.

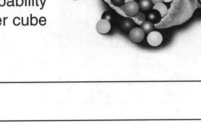

_____

_____

_____

_____

APPLY YOUR SKILLS:

In a standard deck of playing cards, there are 13 hearts, 13 diamonds, 13 spades, and 13 clubs. The deck is shuffled and you are dealt one card. Develop a probability model that could be used to determine the probability of being dealt a club. Explain how you determined your answer.

_____

_____

_____

_____

_____

Name: _____ Date: _____

STATISTICS AND PROBABILITY – Develop Probability Models

CCSS Math Content 7.SP.C.7b: Develop a probability model (which may not be uniform) by observing frequencies in data generated from a chance process.

SHARPEN YOUR SKILLS:

Luke has 48 colored bouncy balls in a bag. He wants Arthur to determine the probability of selecting a green ball. Arthur draws a ball from the bag and records its color in a table. He then returns the ball to the bag. Luke shakes the bag up to mix the balls. Then Arthur draws and records the color of another ball. The boys repeat this process 20 times. The results of their experiment are shown in the table.

Color of Ball	Frequency
Red	4
Yellow	6
Green	10

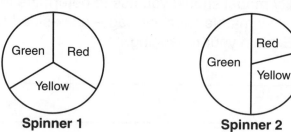

Spinner 1   Spinner 2

Based on the results of the experiment, which spinner should Arthur use to determine the probability of drawing a green ball? Explain your reasoning.

_____

_____

_____

_____

APPLY YOUR SKILLS:

The names of 100 people are written on slips of paper and placed in a bag. Names of females are written on yellow paper and names of males are written on green paper. A slip of paper is drawn, and the gender of the person is recorded. The paper is returned to the bag, and the bag is shaken. This process is repeated 12 times. The results of the experiment are shown in the table. Use the results of the experiment to develop a model that could be used to determine the probability of drawing a female name. Explain how you determined your answer.

Gender	Frequency
Female	3
Male	9

_____

_____

_____

_____

_____

_____

Name: _____ Date: _____

STATISTICS AND PROBABILITY –
Understanding Compound Probabilities

CCSS Math Content 7.SP.C.8a: Understand that, just as with simple events, the probability of a compound event is the fraction of outcomes in the sample space for which the compound event occurs.

SHARPEN YOUR SKILLS:

Ling flips two fair coins. What is the probability that both coins land on heads? Explain how you determined your answer.

_____

_____

_____

APPLY YOUR SKILLS:

Mr. Vogel flips a fair coin and rolls a six-sided number cube. He asks his students to determine the probability that the coin lands on tails and the number cube lands on an even number.

Student 1 says the probability is $\frac{1}{2}$, because the coin has only two different outcomes.

Student 2 says the probability is $\frac{1}{2}$, because the coin has only two different outcomes and half of the numbers on a number cube are even numbers.

Student 3 says the probability is $\frac{1}{4}$, because the coin has only two different outcomes and half of the numbers on a number cube are even numbers.

Which student is correct? Explain how you determined your answer.

_____

_____

_____

_____

_____

_____

Name: _____ Date: _____

STATISTICS AND PROBABILITY –
Representing Sample Spaces

CCSS Math Content 7.SP.C.8b: Represent sample spaces for compound events using methods such as organized lists, tables, and tree diagrams. For an event described in everyday language (e.g., "rolling double sixes"), identify the outcomes in the sample space which compose the event.

SHARPEN YOUR SKILLS:

Use an organized list or table to represent the sample space for rolling two dice. Event *A* is defined as all outcomes whose sum is greater than or equal to 8. Circle all outcomes in your sample space that are part of Event *A*.

APPLY YOUR SKILLS:

Three fair coins are tossed. Event *B* is defined as outcomes in which there are at least two heads. Lea argues that there are 4 outcomes in Event *B*, and Amanda argues that there are only 3. Who is correct? Explain how you determined your answer.

_____

_____

_____

_____

_____

_____

Name: _____ Date: _____

STATISTICS AND PROBABILITY –
Representing Sample Spaces

CCSS Math Content 7.SP.C.8b: Represent sample spaces for compound events using methods such as organized lists, tables, and tree diagrams. For an event described in everyday language (e.g., "rolling double sixes"), identify the outcomes in the sample space which compose the event.

SHARPEN YOUR SKILLS:

Justin has three pairs of pants (black, brown, and blue) and four shirts (white, red, green, and purple). Event *C* is defined as the outfits that have brown pants. Draw a tree diagram to show how many outfits Justin can make. Then list the outcomes in Event *C*.

_____

_____

_____

APPLY YOUR SKILLS:

Lloyd works at a sandwich shop. The shop has three types of bread (wheat, white, and gluten free), three types of meat (turkey, ham, and roast beef), and three types of cheese (cheddar, Swiss, and provolone). Using one type of bread, one type of meat, and one type of cheese, how many different sandwiches can Lloyd make? How many of these sandwiches are on white bread? On your own paper, draw a tree diagram to organize the sample space and support your answer.

_____

_____

Name: _____   Date: _____

STATISTICS AND PROBABILITY –
Designing and Simulating Compound Events

CCSS Math Content 7.SP.C.8c: Design and use a simulation to generate frequencies for compound events.

SHARPEN YOUR SKILLS:

Research shows that 10% of people are left handed. Donovan wants to survey left-handed people for a research project. He has a software program that will randomly select the name of someone in his city and provide a phone number for that person. Donovan wants to determine how many people he would have to call in order to survey 5 left-handed people. Let the number 1 represent a left-handed person. Use a random number generator on a calculator or computer with the numbers 0 through 9 to simulate this situation. Then, based on the simulation results, write a sentence explaining how many people Donavan will have to call in order to survey 5 left-handed people.

_____

_____

_____

_____

_____

APPLY YOUR SKILLS:

Write a short paragraph explaining why the simulation above works for Donovan's situation.

_____

_____

_____

_____

_____

_____

Name: _____ Date: _____

STATISTICS AND PROBABILITY –
Designing and Simulating Compound Events

CCSS Math Content 7.SP.C.8c: Design and use a simulation to generate frequencies for compound events.

SHARPEN YOUR SKILLS:

Approximately 33% of people are lactose intolerant. An ice cream shop in the mall randomly selects shoppers to taste their newest flavor of ice cream and complete a survey about it. Design and use a simulation to determine the number of people who will be randomly selected to participate in the ice cream survey before 3 people who are lactose intolerant are selected.

_____

_____

_____

_____

_____

_____

APPLY YOUR SKILLS:

Write a short paragraph explaining why the simulation you designed above works for the ice cream shop situation.

_____

_____

_____

_____

_____

Answer Keys

GEOMETRY

Scale Drawings (pg. 1)
SHARPEN YOUR SKILLS:
Length of Shen's Backyard:

$$8 \text{ ft/cm} \times 10\frac{1}{2} \text{ cm} = \frac{8 \text{ ft}}{1 \text{ cm}} \times \frac{21 \text{ cm}}{2} = \frac{168 \text{ ft}}{2} = 84 \text{ ft}$$

Width of Shen's Backyard:

$$8 \text{ ft/cm} \times 8\frac{3}{4} \text{ cm} = \frac{8 \text{ ft}}{1 \text{ cm}} \times \frac{35 \text{ cm}}{4} = \frac{280 \text{ ft}}{4} = 70 \text{ ft}$$

Actual Area of Shen's Backyard:
$A = l \times w$ 84 ft $\times$ 70 ft = 5,880 ft^2
The actual area of Shen's backyard is 5,880 square feet.

APPLY YOUR SKILLS:
Length of Scale Drawing of Shen's Backyard:

$$\frac{\frac{1}{2} \text{ in.}}{7 \text{ ft}} \times \frac{84 \text{ ft}}{1} = \frac{42 \text{ in.}}{7} = 6 \text{ in.}$$

Width of Scale Drawing of Shen's Backyard:

$$\frac{\frac{1}{2} \text{ in.}}{7 \text{ ft}} \times \frac{70 \text{ ft}}{1} = \frac{35 \text{ in.}}{7} = 5 \text{ in.}$$

The scale drawing should be 6 inches by 5 inches.

Drawing and Constructing Geometric Shapes (pg. 2)
SHARPEN YOUR SKILLS:
The triangles that students draw will vary because they may choose different side lengths. However, all of the triangles should have one angle that measures 21°, one that measures 42°, and one that measures 117°.

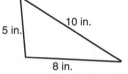

APPLY YOUR SKILLS:
1. This statement is *never* true. If given three angles whose sum is 180°, you can draw many different triangles, as the size of the angle has no affect on the length of the sides. For example, the triangles below have angles with the same measures, but their sides are different lengths.

2. This statement is *always* true. If given three angles whose sum is 180°, you can draw many different triangles, as the size of the angle has no affect on the length of the sides. For example, the triangles below have angles with the same measures, but their sides are different lengths.

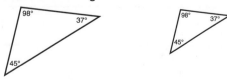

3. This statement is *always* true. If the sum of the measures of the three angles is less than or greater than 180°, then no triangle can be drawn using those three angles. For example, the figures below show that if the sum of the angles is greater than 180° or less than 180°, no triangle is formed.

4. This statement is *always* true. If the sum of the lengths of any two sides is greater than the length of the third side, then one unique triangle can be drawn. For example, in the triangle below, the sum of any two sides is greater than the length of the third side.

5. This statement is *never* true. If the sum of the lengths of any two sides is greater than the length of the third side, then one unique triangle can be drawn. For example, the triangles below have the same side lengths. Although they have different orientations, they are congruent.

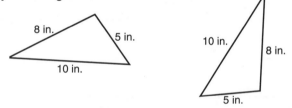

6. This statement is *always* true. If the sum of the lengths of any two sides is less than the length of the third side, then no triangle can be drawn. For example, the sum of 7 centimeters and 9 centimeters is 16 centimeters, which is less than 19 centimeters. The figure below shows that a triangle cannot be formed with these side lengths.

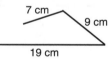

Cross-Sections of Solids (pg. 3)
SHARPEN YOUR SKILLS:

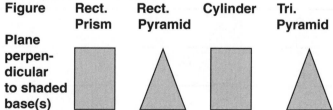

Figure	Rect. Prism	Rect. Pyramid	Cylinder	Tri. Pyramid
Plane perpendicular to shaded base(s)				

Figure	Rect. Prism	Rect. Pyramid	Cylinder	Tri. Pyramid
Plane parallel to shaded base(s)	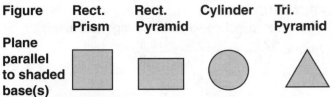			

APPLY YOUR SKILLS:

1. Yes; All of the cross-sections of a right rectangular prism will be congruent, because the sides of the prism are perpendicular to the bases.

2. No; The cross-sections of a right rectangular pyramid will not be congruent, because the sides of the pyramid are *not* perpendicular to the base. The cross-sections become smaller as the planes slicing it get closer to the apex.

Circumference and Area of Circles (pg. 4)
SHARPEN YOUR SKILLS:

1. $C = 2\pi r$ $A = \pi r^2$
   $= 2\pi(4) = 8\pi$ ft $= \pi(4^2) = 16\pi$ ft^2
   The circumference is 8π feet, and the area is 16π square feet.

2. $C = \pi d$ $A = \pi r^2$
   $= \pi(13) = 13\pi$ m $= \pi(6.5)^2 = 42.25\pi$ m^2
   The circumference is 13π meters and the area is 42.25π square meters.

APPLY YOUR SKILLS:

$A = \pi r^2$
   $= \pi(12)^2 = 144\pi$ ft$^2 \approx 452.16$ ft^2
The area of Albert's patio is approximately 452.16 square feet.

$$452.16 \text{ ft}^2 \times \frac{4.5 \text{ bricks}}{1 \text{ ft}^2} \approx \frac{452.16 \text{ ft}^2}{1} \times \frac{4.5 \text{ bricks}}{1 \text{ ft}^2}$$
$$= 2{,}034.72 \text{ bricks}$$
$$\approx 2{,}035 \text{ bricks}$$

It will take approximately 2,035 bricks to make Albert's patio.

Angle Relationships (pg. 5)
SHARPEN YOUR SKILLS:

1. $3a - 5 + 2a = 90$
   $5a - 5 = 90$
   $5a = 95$
   $a = 19$

   Measure of $\angle A$ Measure of $\angle B$
   $3a - 5 = 3(19) - 5$ $2a = 2(19)$
   $= 57 - 5$ $= 38$
   $= 52$

   The measure of $\angle A$ is 52°, and the measure of $\angle B$ is 38°.

2. $24 - 5y - 7y = 180$
   $24 - 12y = 180$
   $-12y = 156$
   $y = -13$

Measure of $\angle C$ Measure of $\angle D$
$24 - 5y = 24 - 5(-13)$ $-7y = -7(-13)$
$= 24 + 65$ $= 91$
$= 89$

The measure of $\angle C$ is 89°, and the measure of $\angle D$ is 91°.

APPLY YOUR SKILLS:

$2x + 4 + 136 = 180$
$2x + 140 = 180$
$2x = 40$
$x = 20$

Measure of $\angle QRS$ Measure of $\angle SRT$
$2x + 4 = 2(20) + 4$ $90 - 44 = 46$
$= 40 = 4$
$= 44$

The measure of $\angle SRT$ is 46°.

Angle Relationships (pg. 6)
SHARPEN YOUR SKILLS:

1. Because $\angle ANG$ and $\angle LNE$ are vertical angles, they have the same measure.
   $8x - 6 = 5x + 12$
   $3x = 18$
   $x = 6$

   The measure of $\angle LNE$ is $8x - 6 = 8(6) - 6$ or 42°. Then, the measure of $\angle ANG$ is also 42°.

2. Angles *SEI* and *IED* are adjacent angles that form $\angle SED$, therefore the sum of the measures is 156°.
   $48 - 2y + 10y + 44 = 156$
   $8y + 92 = 156$
   $8y = 64$
   $y = 8$

   The measure of $\angle SEI$ is $48 - 2y = 48 - 2(8)$ or 32°. Then, the measure of $\angle IED$ is $156° - 32°$ or 124°.

APPLY YOUR SKILLS:

Because angles *MEA* and *MER* are adjacent angles that form a straight line, I know that their measures must have a sum of 180°.

$5a - 9 + 7a + 33 = 180$
$12a + 24 = 180$
$12a = 156$
$a = 13$

The measure of $\angle MER$ is $7a + 33 = 7(13) + 33$ or 124°. Because $\angle MER$ and $\angle AEU$ are vertical angles, I know that they have the same measure. So, the measure of $\angle AEU$ is 124°. Because $\angle AES$ and $\angle SEU$ are adjacent angles that form $\angle AEU$, I know that the sum of their measures is 124°.

$85 - 3b + 2b + 42 = 124$
$127 - b = 124$
$-b = -3$
$b = 3$

Then, the measure of $\angle SEU$ is $2b + 42 = 2(3) + 42$ or 48°.

Problem Solving With Area, Volume, and Surface Area (pg. 7)

SHARPEN YOUR SKILLS:

1. Area of isosceles triangle

$A = \frac{1}{2}bh$

$= \frac{1}{2}(7.2)(4.8)$

$= \frac{1}{2}(34.56)$

$= 17.28 \text{ m}^2$

Area of parallelogram

$A = bh$

$= \left(7\frac{1}{2}\right)(6)$

$= \left(\frac{15}{2}\right)\left(\frac{6}{1}\right)$

$= \frac{90}{2}$

$= 45 \text{ yd}^2$

2. Area of one base

$A = \frac{1}{2}(b_1 + b_2)h$

$= \frac{1}{2}(22 + 16)(19)$

$= \frac{1}{2}(38)(19)$

$= (19)(19)$

$= 361 \text{ ft}^2$

Lateral area

$A = Ph$

$= (22 + 16 + 20.5 + 20.5)(29)$

$= (79)(29)$

$= 2,291 \text{ ft}^2$

Surface area

$SA = 2B + LA$

$= 2(361) + 2291$

$= 722 + 2291$

$= 3,013 \text{ ft}^2$

The surface area of the prism is 3,013 square feet.

APPLY YOUR SKILLS:

Area of top of cake

$A = bh$

$= (12)(10)$

$= 120 \text{ in.}^2$

Lateral area of cake

$A = Ph$

$= (10 + 12 + 10 + 12)\left(3\frac{1}{4}\right)$

$= (44)\left(3\frac{1}{4}\right)$

$= 143 \text{ in.}^2$

The total amount of cake that needs to be covered with fondant is 120 in.² + 143 in.² or 263 square inches. Meredith will have to make two batches of fondant to cover the cake. She will then have enough fondant to cover 500 square inches of cake. So, she will *not* use all of the fondant.

Problem Solving With Area, Volume, and Surface Area (pg. 8)

SHARPEN YOUR SKILLS:

1. Volume of the cube

$V = s^3$

$= \left(8\frac{5}{8}\right)^3$

$= \left(\frac{69}{8}\right)^3$

$= \frac{328,509}{512}$

$= 641\frac{317}{512} \text{ ft}^3$

2. Volume of triangular prism

$V = Bh$

$= \left[\frac{1}{2}(30)(36)\right](61)$

$= \left[\frac{1}{2}(1080)\right](61)$

$= (540)(61)$

$= 32,940 \text{ cm}^3$

APPLY YOUR SKILLS:

Convert feet to inches—Plastic storage container:

$1.75 \text{ ft} \times \frac{12 \text{ in.}}{1 \text{ ft}} = \frac{1.75 \text{ ft}}{1} \times \frac{12 \text{ in.}}{1 \text{ ft}}$

$= \frac{21 \text{ in.}}{1}$

$= 21 \text{ in.}$

Volume of plastic storage container:

$V = Bh$

$= (16)(13)(21)$

$= 4,368 \text{ in.}^3$

Convert feet to inches—Box of clothes:

$1.5 \text{ ft.} \times \frac{12 \text{ in.}}{1 \text{ ft}} = \frac{1.5 \text{ ft}}{1} \times \frac{12 \text{ in.}}{1 \text{ ft}}$

$= \frac{18 \text{ in.}}{1}$

$= 18 \text{ in.}$

Volume of box of clothes:

$V = Bh$

$= (18)(18)(14)$

$= 4,536 \text{ in.}^3$

The volume of the box is greater than the volume of the plastic storage container. So, Devon cannot fit all of the clothes currently in the box into the plastic storage container.

RATIOS AND PROPORTIONAL RELATIONSHIPS

Unit Rates (pg. 9)

SHARPEN YOUR SKILLS:

1. $\dfrac{2 \text{ tsp}}{\frac{3}{4} \text{ c}} = \dfrac{x}{1 \text{ c}}$

$\left(\frac{3}{4}\cancel{c}\right)(x) = (2 \text{ tsp})(1\cancel{c})$

$\frac{3}{4}x = 2 \text{ tsp}$

$x = \left(\frac{2 \text{ tsp}}{1}\right)\left(\frac{4}{3}\right)$

$x = \frac{8}{3} \text{ tsp}$

$x = 2\frac{2}{3} \text{ tsp}$

The unit rate is $2\frac{2}{3}$ teaspoons of vanilla to 1 cup of sugar.

2. $\dfrac{\frac{3}{20} \text{ marathon}}{\frac{2}{3} \text{ hour}} = \dfrac{x}{1 \text{ hour}}$

$\left(\frac{2}{3}\cancel{\text{hour}}\right)(x) = \left(\frac{3}{20} \text{ marathon}\right)(1\cancel{\text{hour}})$

$\frac{2}{3}x = \frac{3}{20} \text{ marathon}$

$x = \left(\frac{3}{20} \text{ marathon}\right)\left(\frac{3}{2}\right)$

$x = \frac{9}{40} \text{ marathon}$

The unit rate is $\frac{9}{40}$ of a marathon in 1 hour.

APPLY YOUR SKILLS:

$$\frac{\frac{1}{4}\text{c}}{\frac{3}{8}\text{oz.}} = \frac{x}{1\text{ oz.}}$$

$$\left(\tfrac{3}{8}\text{oz.}\right)(x) = \left(\tfrac{1}{4}\text{c}\right)(1\text{ oz.})$$

$$\tfrac{3}{8}x = \tfrac{1}{4}\text{c}$$

$$x = \left(\tfrac{1}{4}\text{c}\right)\left(\tfrac{8}{3}\right)$$

$$x = \tfrac{8}{12}\text{c}$$

$$x = \tfrac{2}{3}\text{c}$$

Eric needs a total of $\frac{2}{3}$ cup of olive oil. So, he must add $\frac{2}{3} - \frac{1}{4}$ or $\frac{5}{12}$ cup of olive oil to the dough.

Proportional Relationships (pg. 10)
SHARPEN YOUR SKILLS:
1. The relationship between x and y is proportional because the ratios of y to x are equivalent. That is, $\frac{7}{1} = \frac{14}{2} = \frac{21}{3} = \frac{28}{4}$.
2. The relationship between x and y is not proportional because the ratios of y to x are not equivalent. That is, $\frac{1}{1} \neq \frac{4}{2} \neq \frac{9}{3} \neq \frac{16}{4}$.
3. The relationship between x and y is proportional because the ratios of y to x are equivalent. That is, $\frac{-1}{-2} = \frac{1}{2} = \frac{2}{4} = \frac{3}{6}$.

APPLY YOUR SKILLS:
1. Answers will vary.
2. According to the values in the table, these quantities are not proportional because the ratios of time to number of laps are not equivalent. That is, $\frac{1}{1} \neq \frac{1.9}{2} \neq \frac{6.5}{5} \neq \frac{12.2}{10}$.

Proportional Relationships (pg. 11)
SHARPEN YOUR SKILLS:
1. This graph does not show a proportional relationship. Although the graph is a straight line, it does not pass through the origin.
2. This graph does not show a proportional relationship. Although the graph passes through the origin, it is not a straight line.
3. This graph does show a proportional relationship. It is a straight line that passes through the origin.

APPLY YOUR SKILLS:
1. The relationship between the price of the stock and time is not proportional because the graph is not a straight line and it does not pass through the origin.
2. The price of the stock is increasing by $5 each year between 2009 and 2013. In order for this relationship to be proportional through 2014, the price of the stock would need to be $70 in 2014.

Constant of Proportionality (pg. 12)
SHARPEN YOUR SKILLS:
1. The value of the y quantity is 8 times the value of the x quantity. Therefore, the constant of proportionality is 8.
2. The value of the y quantity changes by 2 as the value of the x quantity changes by 7. Therefore, the constant of proportionality is $\frac{2}{7}$.

APPLY YOUR SKILLS:
As the number of hours increases by 1, the amount Maya charges increases by $7.50. Therefore, Maya charges $7.50 per hour.
$$\frac{\$15}{2h} = \frac{\$7.50}{1h}$$

Constant of Proportionality (pg. 13)
SHARPEN YOUR SKILLS:
1. The constant of proportionality is 2.
2. The constant of proportionality is $\frac{5}{3}$.
3. $\frac{\$4.80}{5\text{ lb}} = \frac{\$0.96}{1\text{ lb}}$
   The price for one pound of apples is $0.96.
4. $\frac{35,000,000 \text{ crayons}}{7 \text{ days}} = \frac{5,000,000 \text{ crayons}}{1 \text{ day}}$
   The company can make 5 million crayons in one day.

APPLY YOUR SKILLS:
To determine which bag of grapes is a better deal, I need to calculate the unit rate for each type.
White grapes: $\frac{\$4.70}{2.5\text{ lb}} = \frac{\$1.88}{1\text{ lb}}$
Red grapes: $\frac{\$5.98}{3.25\text{ lb}} = \frac{\$1.84}{1\text{ lb}}$
Because the unit rate for the red grapes ($1.84 per pound) is less than the unit rate for the white grapes ($1.88 per pound), the red grapes are a better deal.

Proportional Relationships (pg. 14)
SHARPEN YOUR SKILLS:
1. A: $f = 15t$ B: $f = 13t$ C: $f = 18t$
   D: $f = 19t$ E: $f = 12t$ F: $f = 14t$
2. A: $f = 14t$ B: $f = 13t$ C: $f = 16t$
   D: $f = 15t$ E: $f = \frac{25}{2}t$ or $2f = 25t$
   F: $f = \frac{29}{2}t$ or $2f = 29t$

APPLY YOUR SKILLS:
The ratio of raisins to sunflower seeds is 2:3 or $\frac{2}{3}$.
So, $\frac{r}{s} = \frac{2}{3}$, where r represents the number of cups of raisins and s represents the number of cups of sunflower seeds. If I solve the equation for s, I get $s = \frac{3}{2}r$. If I solve the equation for r, I get $r = \frac{2}{3}s$. Therefore, the equations represent the same relationship.

Proportional Relationships and Graphs (pg. 15)
SHARPEN YOUR SKILLS:
1. Point *A* represents the distance Darrell rides his bicycle in one hour. So, Darrell rides his bicycle at a rate of 14 miles per hour.
2. The point (0, 0) represents the distance Darrell rides his bicycle in 0 hours. So, Darrell rides his bicycle 0 miles in 0 hours.

APPLY YOUR SKILLS:
Luther sells the corn at a rate of $5.28 for 3 dozen or $1.76 per dozen. Point *B* corresponds to the rate at which Luther sells the corn, because 4.5 dozen ears of corn at $1.76 per dozen is $7.92.

Problem Solving With Proportional Relationships (pg. 16)
SHARPEN YOUR SKILLS:
Change in asking price: $372,000 – $327,360 = $44,640
Percent decrease in asking price:

$$\frac{44,640}{372,000} = \frac{x}{100}$$

$$(44,640)(100) = 372,000x$$

$$\frac{4,464,000}{372,000} = \frac{372,000x}{372,000}$$

$$12 = x$$

There has been a 12% decrease in the asking price of Margaret's house.

APPLY YOUR SKILLS:

Packs of Baseball Cards	Cost With Tax	Number of Hats	Cost With Tax	Total Cost With Tax
6	$17.74	1	$32.20	$49.94
12	$35.48	2	$64.39	$99.87
18	$53.21	3	$96.59	$149.80
24	$70.95	4	$128.79	$199.74

River can buy 24 packs of baseball cards and 4 baseball hats.

THE NUMBER SYSTEM

Opposite Quantities With Zero Sum (pg. 17)
SHARPEN YOUR SKILLS:
1. This *is not* a situation in which opposite quantities combine to make 0, because –$12.84 and $20.00 are not opposite quantities.
2. This *is* a situation in which opposite quantities combine to make 0, because a loss of 10 yards (or –10 yard gain) and a gain of 10 yards are opposite quantities.
3. This *is not* a situation in which opposite quantities combine to make 0, because an elevation of 345 feet above sea level and 500 feet below (or –500 feet) sea level are not opposite quantities.

4. This *is* a situation in which opposite quantities combine to make 0, because a balance of $200 and a withdrawal of $200 (or –$200) are opposite quantities.

APPLY YOUR SKILLS:
Answers will vary.

Understanding Addition of Rational Numbers (pg. 18)
SHARPEN YOUR SKILLS:
1. Point *D* represents the sum 4 + (–5). It indicates the number that is a distance of 5 from 4 in the negative direction, which is –1.
2. Point *J* represents the sum 4 + 5. It indicates the number that is a distance of 5 from 4 in the positive direction, which is 9.
3. Point *F* represents the sum –4 + 5. It indicates the number that is a distance of 5 from –4 in the positive direction, which is 1.
4. Point *A* represents the sum –4 + (–5). It indicates the number that is a distance of 5 from –4 in the negative direction, which is –9.
5. Point *E* represents the sum 4 + (–4). It indicates the number that is a distance of 4 from 4 in the negative direction, which is 0.

APPLY YOUR SKILLS:
1. –8 + 6
2. The temperature at noon is 6 away from –8 in the positive direction, which is –2. Therefore, the temperature at noon was –2°F.

Understanding Subtraction of Rational Numbers (pg. 19)
SHARPEN YOUR SKILLS:
1. 8 – 3 = 8 + (–3)
2. 9.2 – (–4.9) = 9.2 + 4.9
3. $-\frac{3}{5} - \frac{2}{5} = -\frac{3}{5} + \left(-\frac{2}{5}\right)$
4. –2 – 7 = –2 + (–7)
5. 4 – 12.3 = 4 + (–12.3)
6. $\frac{2}{9} - \frac{5}{6} = \frac{2}{9} + \left(-\frac{5}{6}\right)$

APPLY YOUR SKILLS:
I plotted both temperatures on a number line diagram.

Looking at the diagram, I can see that the distance between –2 and –9 is 7. Therefore, the change in temperature is |–2 – (–9)| or 7 degrees.

Adding and Subtracting Rational Numbers (pg. 20)
SHARPEN YOUR SKILLS:
1. 15 + (–47) = –32
2. $-\frac{4}{7} + \frac{6}{7} = \frac{2}{7}$
3. –28.73 + (–42.21) = –70.94
4. $\frac{4}{15} + \frac{7}{20} = \frac{16}{60} + \frac{21}{60} = \frac{37}{60}$

APPLY YOUR SKILLS:

1. $5.3 + 22\frac{1}{2} = 5.3 + 22.5 = 27.8$
   Danielle will work a total of 27.8 hours this week.
2. $-2.5 + 7.4 = 4.9$
   The tree is 4.9 feet tall above ground.

Adding and Subtracting Rational Numbers (pg. 21)
SHARPEN YOUR SKILLS:

1. $13.7 - 86.9 = -73.2$
2. $-\frac{5}{9} - \left(-\frac{8}{9}\right) = \frac{3}{9}$ or $\frac{1}{3}$
3. $58 - 17\frac{3}{8} = 57\frac{8}{8} - 17\frac{3}{8} = 40\frac{5}{8}$
4. $-161 - 122 = -283$

APPLY YOUR SKILLS:

1. $1.88 - 1\frac{7}{10} = 1.88 - 1.7 = 0.18$
   Roderick is 0.18 meters or 18 centimeters taller than Jacqueline.
2. $17 - (-9) = 26$
   Sarah traveled 26 floors.

Understanding Multiplication of Rational Numbers (pg. 22)
SHARPEN YOUR SKILLS:

For this section, answers may vary. Sample answers are shown.

1. $\left(\frac{2}{5} \times \left(-\frac{5}{8}\right)\right) \times \left(-\frac{4}{7}\right) = \frac{2}{5} \times \left(\left(-\frac{5}{8}\right) \times \left(-\frac{4}{7}\right)\right)$
   $= \frac{2}{5} \times \left(\frac{5}{8} \times \frac{4}{7}\right)$

2. $-8((-1.5) + 4.25) = (-8)(-1.5) + (-8)(4.25)$
   $= (8)(1.5) + (-8)(4.25)$

APPLY YOUR SKILLS:

I can represent the change in Christopher's elevation each hour with the negative rational number $-\frac{4}{5}$. Then, to determine the total change in his elevation over the course of Christopher's hike, I would multiply this by 3. Therefore, the change in Christopher's elevation can be determined using the expression $\left(-\frac{4}{5}\right)(3)$.

Understanding Division of Rational Numbers (pg. 23)
SHARPEN YOUR SKILLS:

1. $-\left(\frac{9}{7}\right) = \frac{(-9)}{7} = \frac{9}{(-7)}$
2. $\frac{(-5)}{8} = -\left(\frac{5}{8}\right) = \frac{5}{(-8)}$
3. $\frac{12}{(-5)} = -\left(\frac{12}{5}\right) = \frac{(-12)}{5}$

APPLY YOUR SKILLS:

I can represent the amount that Tim's company lost over the last 6 months with the negative integer −32,857. Then, to estimate how much money was lost each month, I would divide by 6. Therefore, the amount that Tim's company lost each month can be estimated by using the expression $\frac{(-32,857)}{6}$.

Multiplying and Dividing Rational Numbers (pg. 24)
SHARPEN YOUR SKILLS:

1. $-18 \times (-4) = 72$
2. $-\frac{4}{9} \times \frac{27}{32} = -\frac{108}{288} = -\frac{3}{8}$
3. $2.45 \times (-12.3) = -30.135$
4. $\frac{4}{15} \times \frac{7}{15} = \frac{28}{225}$

APPLY YOUR SKILLS:

1. $-182.25 \times 6\frac{1}{2} = -182.25 \times 6.5 = -1,184.625$
   The submarine will be 1,184.625 feet below sea level after $6\frac{1}{2}$ minutes.
2. $6 \times 37.5 = 225$
   Katie's Pie Place sold 225 pieces of pie last week.

Multiplying and Dividing Rational Numbers (pg. 25)
SHARPEN YOUR SKILLS:

1. $\frac{4}{5} \div \left(-\frac{8}{9}\right) = \frac{4}{5} \times \left(-\frac{9}{8}\right) = -\frac{36}{40}$ or $-\frac{9}{10}$
2. $-475 \div (-19) = 25$
3. $-\frac{18}{35} \div \frac{3}{7} = -\frac{18}{35} \times \frac{7}{3} = -\frac{126}{105}$ or $-\frac{6}{5}$ or $-1\frac{1}{5}$
4. $314.55 \div (-13.98) = -22.5$

APPLY YOUR SKILLS:

1. $-3,280.6 \div 500 = -6.5612$
   Therefore, the elevator will make 6 stops before reaching its greatest depth.
2. $94,820 \div 6 \approx 15,803.\overline{33}$
   The mining company brought up approximately 15,803.33 tons of coal per day.

Writing Rational Numbers as Decimals (pg. 26)
SHARPEN YOUR SKILLS:

1.
```
        0.681̄8̄1̄
   22 )15.00000
      − 132
        180
      − 176
         40
       − 22
        180
      − 176
         40
       − 22
         18
```

2.
```
       0.625
    8 )5.000
      − 48
        20
      − 16
        40
      − 40
         0
```

3.
```
         1.286
   500 )643.000
       − 500
        1430
       − 1000
        4300
       − 4000
        3000
       − 3000
           0
```

APPLY YOUR SKILLS:

Sample answer: A rational number is a number whose decimal form terminates in 0s or eventually repeats. If I use long division to convert $\frac{22}{7}$ to its decimal form, I get $3.\overline{142857}$. Therefore, Eva is incorrect. Although π is not a rational number, $\frac{22}{7}$ is rational.

Problem Solving With Rational Numbers (pg. 27)
SHARPEN YOUR SKILLS:

1. $\frac{3}{16} + \frac{5}{8} = \frac{3}{16} + \frac{10}{16} = \frac{13}{16}$

   The number $\frac{13}{16}$ is $\frac{5}{8}$ more than $\frac{3}{16}$.

2. $\frac{9}{10} - \frac{1}{4} = \frac{18}{20} - \frac{5}{20} = \frac{13}{20}$

   The library is $\frac{13}{20}$ of a mile farther from the school than the community center.

APPLY YOUR SKILLS:

1. $1\frac{3}{4} + \frac{1}{3} = \frac{21}{12} + \frac{4}{12} = \frac{25}{12}$ or $2\frac{1}{12}$

   The curtain will be $2\frac{1}{12}$ yards long after the lace has been added.

2. $\frac{7}{8} - \frac{1}{3} = \frac{21}{24} - \frac{8}{24} = \frac{13}{24}$

   There will be $\frac{13}{24}$ of a yard of lace left.

Problem Solving With Rational Numbers (pg. 28)
SHARPEN YOUR SKILLS:

1. $\frac{9}{14} \times \frac{2}{3} = \frac{18}{42}$ or $\frac{3}{7}$

   The number $\frac{3}{7}$ is $\frac{2}{3}$ of $\frac{9}{14}$.

2. $\frac{3}{4} \div \frac{1}{8} = \frac{3}{4} \times \frac{8}{1} = \frac{24}{4}$ or 6

   Steven will have 6 piles of topsoil.

APPLY YOUR SKILLS:

1. $18\frac{1}{2} \times 12\frac{1}{4} = \frac{37}{2} \times \frac{49}{4} = \frac{1813}{8}$ or $226\frac{5}{8}$

   DeWayne's garden is $226\frac{5}{8}$ square feet.

2. Area of each rectangle:

   $3\frac{5}{8} \times 1\frac{2}{3} = \frac{29}{8} \times \frac{5}{3} = \frac{145}{24}$ or $6\frac{1}{24}$

   The area of each rectangle is $6\frac{1}{24}$ square feet.

   $226\frac{5}{8} \div 6\frac{1}{24} = \frac{1813}{8} \div \frac{145}{24} = \frac{1813}{8} \times \frac{24}{145} = \frac{43512}{1160}$ or $37\frac{74}{145}$

   DeWayne will have about 38 rectangles.

EXPRESSIONS AND EQUATIONS

Manipulating Linear Expressions (pg. 29)
SHARPEN YOUR SKILLS:

1. $\left(\frac{2}{7}x + 8\right) + \left(\frac{3}{7}x - 4\right) = \frac{5}{7}x + 4$

2. $\left(\frac{11}{15}a - 21\right) - \left(\frac{2}{5}a + 16\right) =$
   $\frac{11}{15}a - 21 - \frac{6}{15}a - 16 = \frac{5}{15}a - 37$ or $\frac{1}{3}a - 37$

APPLY YOUR SKILLS:

The student made a mistake when removing the parentheses. In the second set of parentheses, the negative was not distributed to both terms. The correct solution is shown.

$\left(\frac{7}{12}y + 4\right) - \left(\frac{5}{6}y - 11\right) + \left(\frac{3}{4}y - 9\right)$

$= \frac{7}{12}y + 4 - \frac{10}{12}y \, \boxed{+} \, 11 + \frac{9}{12}y - 9$

$= \frac{7}{12}y - \frac{10}{12}y + \frac{9}{12}y + 4 + 11 - 9$

$= \frac{6}{12}y + 6$

$= \frac{1}{2}y + 6$

Manipulating Linear Expressions (pg. 30)
SHARPEN YOUR SKILLS:

1. $\frac{5}{24}x + \frac{10}{36} = \frac{5}{12}\left(\frac{1}{2}x + \frac{2}{3}\right)$

2. $\frac{4}{9}\left(\frac{2}{5}b - \frac{3}{8}\right) = \frac{8}{45}b - \frac{12}{72}$

3. $\frac{6}{45} - \frac{8}{63}y = \frac{2}{9}\left(\frac{3}{5} - \frac{4}{7}y\right)$

4. $-\frac{2}{3}\left(\frac{8}{9} - \frac{2}{5}a\right) = -\frac{16}{27} + \frac{4}{15}a$

APPLY YOUR SKILLS:

Yes, there is more than one way to factor the expression $\frac{24}{32}a + \frac{48}{56}$. Two examples are shown below.

$\frac{24}{32}a + \frac{48}{56} = \frac{2}{4}\left(\frac{12}{8}a + \frac{24}{14}\right)$ or $\frac{24}{32}a + \frac{48}{56} = \frac{12}{8}\left(\frac{2}{4}a + \frac{4}{7}\right)$

Rewriting Linear Expressions (pg. 31)
SHARPEN YOUR SKILLS:

If I simplify expression #1, I get $\frac{13}{10}r$ or $1.3r$. Therefore, the expressions are not equivalent. If expression #2 is divided by 10, the resulting expression would be equivalent to expression #1.

APPLY YOUR SKILLS:

I can rewrite the expression $x + 0.08x$ as $1.08x$. This expression makes it easier to calculate the amount that client #2 bid, because I simply have to multiply the amount that client #1 bid by 1.08 to determine the amount that client #2 bid.

Problem Solving With Rational Numbers (pg. 32)
SHARPEN YOUR SKILLS:

Cost $= 0.25\left(7\frac{3}{4}\right) + 1.23 = 0.25(7.75) + 1.23$
$ = 1.9375 + 1.23 = 3.1675$

It will cost Mrs. Teurk about $3.17 for the bulletin board paper.

APPLY YOUR SKILLS:

Sample answer: One and a half years is equivalent to 18 months. Since the calf is already 8 months old, it has 10 months to grow at a rate of 200 pounds per day. There are about 30 days in a month. So, there are approximately 10×30 or 300 days in 10 months.

The blue whale calf will gain about $\frac{200 \text{ lbs.}}{\text{day}}$ x 300 days or 60,000 pounds in 10 months.

Converting this amount to tons, I get 60,000 ÷ 2,000 = 30 tons. So, the calf will weigh 30 + 23 $\frac{3}{8}$ or 53 $\frac{3}{8}$ tons when it is 1.5 years old.

Problem Solving With Equations (pg. 33)
SHARPEN YOUR SKILLS:

1. $\frac{1}{5}x + 7 = 16$

$\frac{1}{5}x = 9$

$x = 45$

2. $2.4y - 9.5 = 4.66$

$2.4y = 14.16$

$y = 5.9$

3. $-\frac{5}{9}a - \frac{2}{3} = \frac{5}{6}$

$-\frac{5}{9}a = \frac{9}{6}$

$a = -\frac{81}{30}$ or $-2\frac{7}{10}$

4. $3.25 + 1.58b = 12.3982$

$1.58b = 9.1482$

$b = 5.79$

APPLY YOUR SKILLS:

Let b represent the number of batches Lois needs to make before she is finished for the day.

$3\frac{1}{4}b + 1.5 = 14\frac{1}{2}$

$3\frac{1}{4}b = 13$

$b = 4$

Lois must make 4 batches of applesauce.

Problem Solving With Equations (pg. 34)
SHARPEN YOUR SKILLS:

1. $\frac{2}{3}(x + 7) = 5$

$x + 7 = \frac{15}{2}$

$x = \frac{1}{2}$

2. $1.9(x - 2.7) = 12.16$

$x - 2.7 = 6.4$

$x = 9.1$

APPLY YOUR SKILLS:

Student #1: $2.5\left(e + 1\frac{1}{2}\right) = 36\frac{7}{8}$

$e + 1\frac{1}{2} = 14.75$

$e = 13.25$

Evan is 13.25 years old.
Student #2:

Evan's Age	Danica's Age	Jeremy's Age	Is Jeremy's Age 36 $\frac{7}{8}$?
10	11.5	28.75	No
15	16.5	41.25	No
13	14.5	36.25	No
13.25	14.75	36.875	Yes

Evan is 13.25 years old.

Both approaches yield a solution of 13.25 years for Evan's age. Student #1's approach is precise and quickly leads to the answer. Student #2's approach requires guessing and checking. While this approach also yielded the correct answer, it is less precise and can be quite time-consuming.

Problem Solving With Inequalities (pg. 35)
SHARPEN YOUR SKILLS:

1. $\frac{8}{15}x + \frac{4}{5} > \frac{29}{30}$

$\frac{8}{15}x > \frac{5}{30}$

$x > \frac{5}{16}$

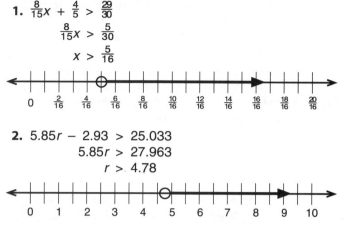

2. $5.85r - 2.93 > 25.033$

$5.85r > 27.963$

$r > 4.78$

APPLY YOUR SKILLS:

Let m be the number of months until Hannah can buy a cell phone.

$85.25m + 482.75 > 1250$

$85.25m > 767.25$

$m > 9$

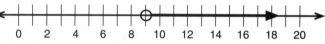

Hannah must save $85.25 for more than 9 months before she can buy a cell phone. So, her account balance will not be more than $1,250.00 until the 10th month.

Problem Solving With Inequalities (pg. 36)
SHARPEN YOUR SKILLS:

1. $15.46r + 28.9 < 240.702$

$15.46r < 211.802$

$r < 13.7$

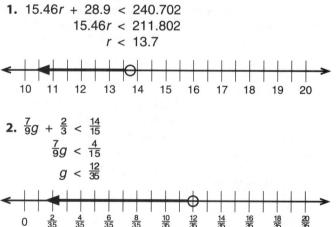

2. $\frac{7}{9}g + \frac{2}{3} < \frac{14}{15}$

$\frac{7}{9}g < \frac{4}{15}$

$g < \frac{12}{35}$

APPLY YOUR SKILLS:

Let h be the number of hours Erin can be away from the beach.

$\frac{19}{25}h + 1\frac{3}{5} < 9\frac{1}{5}$

$\frac{19}{25}h < \frac{38}{5}$

$h < 10$

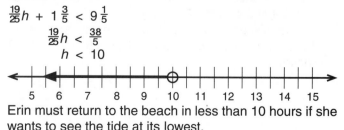

Erin must return to the beach in less than 10 hours if she wants to see the tide at its lowest.

STATISTICS AND PROBABILITY

Using Statistics to Understand Populations (pg. 37)
SHARPEN YOUR SKILLS:

1. This statistic is *not* valid because the sample is not representative of the population. The statistic is only representative of those who attended amusement parks, not of the American population.
2. This statistic is *not* valid because the sample is not representative of the population. The statistic is only representative of the people in the surveyor's 3rd period class.

APPLY YOUR SKILLS:

Sample answer:

For exercise 1, she would have needed to randomly sample all types of Americans, not just those who visited amusement parks.

For exercise 2, he would have needed to take a random sample from all of the seventh graders at the school.

Using Data to Draw Inferences (pg. 38)
SHARPEN YOUR SKILLS:

Sample answer: Based on the data Jake collected, it appears that about 33% of the fish in the pond are largemouth bass and about 67% are bluegills. If there are 120 fish in the pond, then 0.33 x 120 or about 40 fish are largemouth bass and 0.67 x 120 or about 80 fish are bluegills.

APPLY YOUR SKILLS:

Sample answer: Fifteen percent of 8 is 1.2. So, it is important that on average no more than 1 pair of jeans is flawed at each hourly inspection. Fifteen percent of 64 is 9.6. So it is important that no more than 9 pairs of jeans are flawed each day. On Day 1, a total of 9 pairs of jeans were flawed. Although the 2nd and 4th samples had more than 1 flawed pair, the average over the course of the day was under the 15%. So, Darlene did not have to adjust the machines on Day 1. On Day 2, a total of 12 pairs of jeans were flawed over the course of the day. Although samples 4, 5, 6, and 8 had zero flawed pairs, the average over the course of the day exceeded 15%. So, Darlene had to make adjustments on Day 2.

Comparing Data Sets With Similar Variabilities (pg. 39)
SHARPEN YOUR SKILLS:

1. The men's team appears to have the greater average height because all but one of the heights is greater than all of the women's heights.
2. The men's team appears to have greater variability in heights, as their heights range from 68 inches to 80 inches.

APPLY YOUR SKILLS:

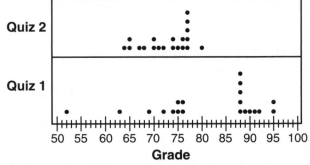

Sample answer: The scores on Quiz 1 likely have a greater average than those on Quiz 2, because over half of the scores on Quiz 1 are 88 or greater. There is more variability in the scores on Quiz 1 as they range from 52 to 95, whereas the Quiz 2 scores range from 64 to 80.

Using Statistics to Infer (pg. 40)
SHARPEN YOUR SKILLS:

1. Neighborhood #1: Mean = $\frac{\$84.62}{10} \approx \$8.46$

   Neighborhood #2: Mean = $\frac{\$104.58}{10} \approx \$10.46$

2. Neighborhood #1: MAD = $\frac{\$7.76}{10} \approx \$0.78$

   Neighborhood #2: MAD = $\frac{\$9.74}{10} \approx \$0.97$

3. Neighborhood #1: Median = $8.20
   Neighborhood #2: Median = $10.30
4. Neighborhood #1: Interquartile Range = $0.95
   Neighborhood #2: Interquartile Range = $2.20

APPLY YOUR SKILLS:

Sample answer: The mean and median for Neighborhood #1 are lower than the mean and median for Neighborhood #2. It appears that, on average, the people who live in Neighborhood #2 earn a higher hourly rate than those in Neighborhood #1. The MAD and interquartile range for Neighborhood #2 are greater than the MAD and interquartile range for Neighborhood #1. This would indicate that there is more variation in the hourly rate earned by those living in Neighborhood #2 than those living in Neighborhood #1.

Understanding Probability (pg. 41)
SHARPEN YOUR SKILLS:

1. 0.1, $\frac{1}{3}$, 0.05; I know that these numbers could represent the probability of an unlikely event because they are close to 0.
2. 0.45, $\frac{7}{15}$, 0.53; I know that these numbers could represent the probability of an event that is neither unlikely nor likely because they are close to $\frac{1}{2}$.

3. $\frac{10}{11}$, 0.99; I know that these numbers could represent the probability of a likely event because they are close to 1.

4. $1\frac{2}{5}$, 3.5; I know that these numbers cannot be probabilities because they are not between 0 and 1.

APPLY YOUR SKILLS:

1. Answers will vary but should be close to 1.

2. Answers will vary but should be close to 0.

3. Answers will vary but should be close to $\frac{1}{2}$.

Using Experimental Probability to Understand Theoretical Probability (pg. 42)
SHARPEN YOUR SKILLS:

1. The probability of rolling a 4 is $\frac{1}{6}$, because there are 6 different numbers on a six-sided number cube, but only 1 of them is a 4.

2. Answers will vary.

3. Answers will vary.

4. Answers will vary.

APPLY YOUR SKILLS:

1. I would expect the coin to land on heads 32 times, because 32 is $\frac{1}{2}$ of 64.

2. I would expect to draw a heart 25 times, because 25 is $\frac{1}{4}$ of 100.

3. I would expect the spinner to land on yellow 30 times, because 30 is $\frac{2}{3}$ of 45.

Uniform Probability Model (pg. 43)
SHARPEN YOUR SKILLS:

I should use the six-sided number cube. There is an equal number of six differently colored marbles in the bag. The probability of drawing a marble of any of those colors out of the bag is equal. Therefore, a six-sided number cube could be used to model this situation.

APPLY YOUR SKILLS:

Answers will vary. However, answers should include a model in which each of four outcomes is equally likely.

Develop Probability Models (pg. 44)
SHARPEN YOUR SKILLS:

Arthur should use Spinner 2 to determine the probability of drawing a green ball. Based on the results of the experiment, there are more green balls than red or yellow balls in the bag. Spinner 2 takes into account the fact that the probabilities are not uniform.

APPLY YOUR SKILLS:

Answers will vary. However, answers should include a model in which the probability of drawing a female name is $\frac{1}{4}$ and the probability of drawing a male name is $\frac{3}{4}$.

Understanding Compound Probabilities (pg. 45)
SHARPEN YOUR SKILLS:

If two coins are flipped, there are four different outcomes: HH, HT, TH, and TT. Only one of those outcomes has two heads. So, the probability that both of Ling's coins will land on heads is $\frac{1}{4}$.

APPLY YOUR SKILLS:

If a coin is flipped and a number cube is rolled, there are 12 different outcomes: H1, H2, H3, H4, H5, H6, T1, T2, T3, T4, T5, and T6. Three of these outcomes have tails and an even number. The probability that the coin will land on tails and the number cube will land on an even number is $\frac{3}{12}$ or $\frac{1}{4}$. So, Student 3 is correct.

Representing Sample Spaces (pg. 46)
SHARPEN YOUR SKILLS:

Sample answer:

1, 1	1, 2	1, 3	1, 4	1, 5	1, 6
2, 1	2, 2	2, 3	2, 4	2, 5	2, 6
3, 1	3, 2	3, 3	3, 4	3, 5	3, 6
4, 1	4, 2	4, 3	4, 4	4, 5	4, 6
5, 1	5, 2	5, 3	5, 4	5, 5	5, 6
6, 1	6, 2	6, 3	6, 4	6, 5	6, 6

The outcomes that are in Event *A* are shaded in the table.

APPLY YOUR SKILLS:

The sample space is: HHH, HHT, HTH, THH, TTT, TTH, THT, HTT. There are 3 outcomes with exactly two heads and 1 outcome with more than two (or three) heads. Therefore, there are a total of 4 outcomes in Event *B*, and Lea is correct.

Representing Sample Spaces (pg. 47)
SHARPEN YOUR SKILLS:

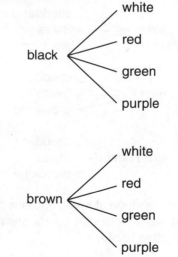

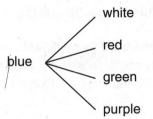

The outcomes that are in Event *C* are: brown pants with white shirt, brown pants with red shirt, brown pants with green shirt, and brown pants with purple shirt.

APPLY YOUR SKILLS:

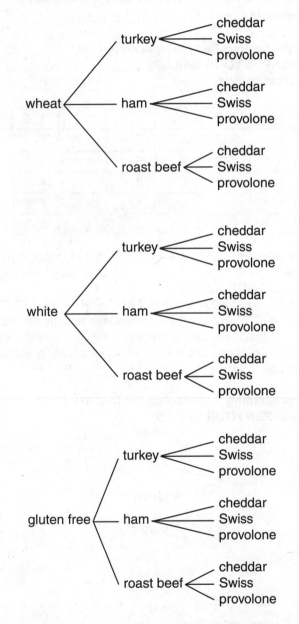

Using one type of bread, one type of meat, and one type of cheese, Lloyd can make 27 different sandwiches. Of these, 9 of them are on white bread.

Designing and Simulating Compound Events (pg. 48)
SHARPEN YOUR SKILLS:
Answers will vary based on individual simulation results, but it would be expected that Donavan will have to call around 50 people in order to survey 5 left-handed people.
APPLY YOUR SKILLS:
There are 10 numbers in 0 – 9. Because 10% of the population is left-handed, designating one of those numbers as being a left-handed person represents 10% of the numbers. Using the random number generator to generate the numbers mimics the random selection of people.

Designing and Simulating Compound Events (pg. 49)
SHARPEN YOUR SKILLS:
Answers will vary based on individual simulation results, but it would be expected that 9 people would need to be surveyed before 3 lactose-intolerant people were selected.
APPLY YOUR SKILLS:
Answers will vary. However, answers should include a simulation in which the probability of surveying a person who is lactose intolerant is approximately 33%.